TECHNICAL CORRESPONDENCE

WILEY SERIES ON HUMAN COMMUNICATION

W. A. Mambert
PRESENTING TECHNICAL IDEAS: A Guide to Audience Communication

William J. Bowman
GRAPHIC COMMUNICATION

Herman M. Weisman
TECHNICAL CORRESPONDENCE: A Handbook and Reference Source for the Technical Professional

John H. Mitchell
WRITING FOR TECHNICAL AND PROFESSIONAL JOURNALS

TECHNICAL CORRESPONDENCE:

A Handbook and Reference Source for the Technical Professional

HERMAN M. WEISMAN
National Bureau of Standards

JOHN WILEY & SONS, INC.
New York · London · Sydney

Sidney Born Technical Library
University of Tulsa

10 9 8 7 6 5 4 3 2

Copyright © 1968 by John Wiley & Sons, Inc.

All rights reserved. This book or any part thereof must not be reproduced in any form without the written permission of the publisher.

Library of Congress Catalog Card Number: 67–30919

GB 471 92640X

Printed in the United States of America

To Margaret, Abbi, Lise, and Harlan

PREFACE

By tradition a preface is supposed to give the author an opportunity to talk directly to his readers to tell them why his book will, at worst, not harm them and, at best, augment their wisdom, add wit and charm to their character, and perhaps brighten, if not their futures, their hopes. I shall make no such claims here for *Technical Correspondence.* I cannot help but agree with Francis Bacon (1561-1626), a great writer of prefaces who candidly admitted that they were a great waste of time and, despite their attempt at modesty, a showy display of boasts and excuses. If most readers are like me, I know they will pass the preface by. They will not have missed anything if they do. If, however, they have been loyal enough to read to this point to find out something about the purpose, objectives, approach, and organization of *Technical Correspondence,* I ask them now to turn to the Introduction. An author has an obligation to explain his book's rationale to his readers. If they find that this book does increase their wisdom, sharpen their wit, magnify their charm, and brighten their hopes and future, that is merely incidental. That potential was within them all the time.

Herman M. Weisman

Washington, D. C.
January, 1968

ACKNOWLEDGMENTS

I wish to make grateful acknowledgment to the following sources for use of material included in this book: American Chemical Society Board of Directors; Joseph D. Becker and *Science;* Board of Editors, *American Scientist; Chemical and Engineering News*; Digital Equipment Corporation, Maynard, Mass.; E. I. du Pont de Nemours & Co., Inc.; Engineering Experiment Station, Iowa State University; Norman A. Evans, Colorado State University; F&M Scientific Division, Hewlett-Packard Company; Fairchild-Davidson, Commack, L. I., New York; Friden, Inc., San Leandro, California; Robert H. Haakenson and Smith Kline & French Laboratories; Thomas T. Holme and *American Scientist*; Jonathon Manufacturing Company, Fullerton, California; Ralph W. Lewis and *Science*; A. S. Marwaha and N. E. Cusack and *Physics Letters,* North-Holland Publishing Company, Amsterdam, The Netherlands; Samuel E. Miller and *Science*; Mark L. Oliphant and *Physics Today*; Harold F. Osborne; Picker Nuclear, White Plains, New York; Milton Rokeach and *Science*; S. A. Rossmassler; Mendel Sachs and *Physics Today*; George H. Soule and *Chemical and Engineering News*; Alfred Weissberg; Howard J. White, Jr.

H. M. W.

CONTENTS

PART I PRINCIPLES AND FUNDAMENTALS

Introduction ... 3

    A Little History ... 4
    Role of Correspondence Today ... 5

Chapter 1. Understanding the Communication Process ... 7

    What Is Communication ... 7
    How Communication Works ... 8
    Meaning ... 10
    Correspondence as a Medium ... 11

Chapter 2. The Psychology, Principles, and Stylistic Practices of Technical Correspondence ... 12

    Psychology in Correspondence ... 12
    Letters Are Personal ... 13
    Know Your Reader ... 15
    Style ... 16
        Clarity ... 18
        Precision ... 18
        Conciseness and Directness ... 19
    Courtesy, Sincerity ... 20
    Unity, Coherence ... 21
    The Problem of Objectivity ... 22

Chapter 3. Planning the Letter ... 24

    Determining Your Purpose ... 27
    Considering Your Reader ... 28

xii Contents

Gathering the Information 28
Outlining and Organizing 28
Writing the Letter 29

Chapter 4. Mechanics of Correspondence 33

Stationery Considerations............................. 33
Framing the Letter 34
Standard Layout Styles of Letters 34
Mechanical Details 40
    Components of the Letter 40
    Heading...................................... 40
    The Date Line 41
    Inside Address 41
    Attention Line, Subject Line, Reference Line 42
    Salutation 44
    Body of the Letter 45
    Complimentary Close 45
    Signature 46
    Identification Line 47
    Enclosures 47
    Carbon Copy Notations 47
    Postscript 48
Second and Succeeding Pages 48
The Envelope 48
The Memorandum 50

PART II APPLYING CORRESPONDENCE PRINCIPLES

Chapter 5. Writing and Replying to Inquiries and Requests 57

The Inquiry Letter 57
Replies to Inquiries and Requests 65
Refusing a Request 78
Order Letters, Acknowledgments, and
    Quotation Letters 80
    Acknowledging Orders 86
    The Quotation Letter 87

Chapter 6. Sales and Proposal Letters 88

Structure of the Sales Letter........................... 90
    The Beginning—Attracting the Reader's Attention 91

Creating Interest in or Desire for the Product or Service	93
Convincing the Reader	94
Motivating the Reader to Act	94
Proposal Letters	98
The Request for Bid	99
The Proposal	101
The Introduction	101
The Technical Presentation	101
The Technical Description	101
Programming	101
Cost Schedules	101
Proposal Transmittal Letters	102
The Letter Proposal	103

Chapter 7. Special-Purpose Letters 106

Complaint and Claims Letters	106
Adjustment Letters	110
The Letter of Instructions	113
The Letter of Authorization	118
The Letter of Transmittal	120

Chapter 8. Employment Letters 122

The View From the Perspective of the Employer	122
Job Analysis and Self-Appraisal	124
The Letter of Application	125
Part One—The Letter	126
Attracting Favorable Attention	126
Adapting Your Qualifications to the Job	129
Securing Action	130
Part Two—The Personal Data Sheet or Resumé	131
Acknowledgment and Follow-Up Letters	141
The Reference Letter	145
The Letter of Recommendation	146
Letters Accepting Positions	149
The Letter of Resignation	149
The Letter Refusing a Job	150

Chapter 9. Personal and Nontechnical Correspondence 152

Letters of Reservation	153
Letters of Introduction	153

Letters of Invitation and Letters of Acceptance or Declination ... 154
    Invitation to Participate in a Committee 155
    An Invitation to Speak 155
    Letters of Acceptance or Declination 157
    Letter of Declination of Appointment 157
Letters of Appreciation and Congratulation 158
Acknowledgments of Letters of
    Appreciation and Congratulations 160
Letters of Condolence 161

Chapter 10. Professional Letters and Memoranda 162

Correspondence Between Professionals 162
Letters to the Editor 166
    The Communications to the Editor 166
    The Editor's Mail Column 169
The Memorandum 176
    The Memorandum For File 179
Meetings 181
    Notices 181
    Agenda for Meetings 182
    Minutes of Meetings 183

PART III   APPENDIX

Index to Grammar, Punctuation, and Usage 193

Selected Bibliography 213

Index 215

PART I

PRINCIPLES AND FUNDAMENTALS

INTRODUCTION

It is fashionable to complain that engineers and scientists cannot write. The complainer can muster many examples and instances in order to build a case. The evidence is real; it contains writing that is diffuse and obscure, frequently disorganized and, more frequently, stereotyped and ungrammatical. At worst, it is unintelligible; at best, it is toilsome reading.

But there are many scientists and engineers who write competently; there are some who write with skillful effectiveness; and a few who write brilliantly.

This book is not addressed to them. It is addressed to the great number of scientists and engineers who meet their communication requirements with difficulty, aversion and, often, with hostility. Some rationalize that skills in communication actually are not part of their professional requirements, hoping that judgment will be based on the technical soundness of their work instead of on their writing skill. Others dismiss skills in communication as arty falderol. As far as they are concerned, there is nothing wrong with (nor incomprehensible about) scientific writing. If it is obscure, it is because the reader is obtuse—unsophisticated and not knowledgeable enough to understand. Furthermore, they gleefully observe that much of the writing in the humanities is far from perfect and often is unintelligible. Others readily admit that they cannot write (or speak) well, but lament that they lack the innate ability to express themselves in words.

There may be some truth in these attitudes—but only just a *little*. The attitudes easily can be refuted. The technical worth of a person's work is evaluated only through communication. Although grammar is certainly not more important than content, its conventions help to reach the audience who must (or should) know about the technically substantive matters worked on. Moreover, even though some persons have more innate ability to manipulate words than others, no particular scientist or engineer is innately unable to communicate. This kind of posture is sheer nonsense. The essence of effective writing is clear thinking. To say that one cannot write is to say that one cannot think. The scientist or engineer, who unabashedly beats his breast to lament that he cannot write, will bridle at the suggestion that he cannot think.

Unfortunately, many scientists and engineers (at some point early in their education) were bitten by one or another species of English teacher—perhaps by both—who was of the genus *Grammarian* or the genus *Literature*. To a teacher of the genus *Grammarian* species, the mechanics of language is next to the word of God. He reduces English to set rules of agreement, punctuation, and spelling. A teacher of the genus *Literature* species is a pedantic prig to whom "good literature" is a cure for faulty metonymy, for insubordinate, coordinate conjunctions, or for lesser evils. His teachings are "out of this world," certainly not of this century; they are frequently beatific raptures and a mishmash of vague aestheticisms. Too few English teachers show their students that the study of English is fundamentally the study of the process of how to communicate something to somebody with the maximum of clarity and effect that is best suited to a particular situation.

The purpose of this book is to examine the process of how to communicate something to somebody with maximum clarity and effect within certain communication situations—not all communication situations but only those encompassed within the title: *Technical Correspondence*. This book is one of a series that examines the process and practices within the full spectrum of communications.

A LITTLE HISTORY

Correspondence has a long, venerable tradition in science and technology. Letters were used by the earliest scientists to exchange information, to report activities and discoveries, to try out ideas, and to receive critical evaluation of experimental results. Frequently (as in the case of Galileo and Kepler), letters were used to justify and defend scientific works. A letter from one scientist was circulated among colleagues at great distances much in the same way as reprints and preprints of publications are circulated today. Galileo, Kepler, Copernicus, and Francis Bacon used correspondence to exchange ideas with scientists in other lands. Business letters and memos helped establish the House of Fugger in fifteenth-century Germany as one of the greatest worldwide commercial empires. Correspondents sent daily letters about German business conditions to all Fugger branches throughout Europe, China, and South America. Fugger agents, in turn, sent intelligence to the main office to report commercial news and the activities of chief competitors. Jacob Fugger organized and analyzed these letters and made his business decisions accordingly.

ROLE OF CORRESPONDENCE TODAY

Today, despite the great capabilities and increasing use of automatic (computer) data and information-processing machinery, correspondence is the basic communication instrument in business and industry just as it was at the start of the industrial revolution 200 years ago. The post office handled almost 80 billion pieces of mail last year. The vast majority of the mail is in the commercial category. The pieces of mail are increasing in excess of 2 billion a year. By 1980, mail volume is expected to exceed 100 billion pieces annually.

Increasingly, automated methods necessarily will be utilized not only to handle the physical pieces of correspondence but also to originate messages of a routine and repetitive nature; however, the growing complexity and specialized nature of our technological society will demand more creative intelligence rather than machines to meet the more complicated and difficult message situations.

Today (and in the foreseeable future), much of the activity of science and technology is and will be conducted through correspondence. It is not only the technical executive or his subordinate manager but also the man in the laboratory or in production who daily must use letters and memoranda. The basic science researcher, as well, in the relative isolation of his laboratory, is called upon to write letters. He sends out inquiries and requests. He answers inquiries and requests. He orders equipment, sends acknowledgments of receipt, or complains when his order is late or faulty. He writes letters of instruction and, on certain occasions, may be called upon to help compose a sales letter or a letter of adjustment. When he wants to change jobs, he will write an application letter. And, on the professional level, he participates in the society that serves his field, using correspondence to further the activity of his organization. He may write a letter to the editor of his society journal, commenting on developments in his profession, criticizing or endorsing views previously expressed in the journal; or he may choose the letter form to communicate through the pages of the periodical of his field on the latest results of his own work and experimentation. Much of the technical man's success in his professional activity depends on his ability to communicate through correspondence. Therefore, it is well for him to be competent in the principles and mechanics of this common (but far from simple) form of communication.

This book is divided into three parts. The first part examines principles and fundamentals. The second part applies the principles to specific correspondence situations. The third part contains an index to grammar, punctuation, and usage, and a selected bibliography.

1
UNDERSTANDING THE COMMUNICATION PROCESS

Samuel Johnson, the eighteenth-century author and dictionary maker, once remarked that he could not understand why anyone wrote except to make money. Implied in his witticism is the lamentably obvious truth that the writing process is so difficult and so painful that, unless a person is amply paid for it, he should not attempt it. Although Johnson's caution was meant, perhaps, for the would-be professional, free-lance writer, his wisdom has wider application. Today, there is not a single professional who, daily, is not called upon to write or to communicate. Competent writing pays off, but writing is difficult. This book is intended to help to make writing easier for the technical professional. (Notice that the adjective used was *easier* not *easy*.)

Yes, writing can be made easier but never easy. Communication is a process through which two or more human beings share each other's thoughts, ideas, feelings, insights, and information, and exchange meanings. As a process, communication is dynamic. Many factors influence the process. Many factors act upon the situation, the communicator, and the communicatee involved. As the process operates, each factor in the communication event acts upon the other factor, and is consequently changed and changing in the process. Because these explanatory words may sound like double talk, let us pause and examine communication. An understanding of it will help make the writing process easier.

WHAT IS COMMUNICATION?

Communication is the way that humans, animals, and insects—even plants—exchange information. Biologists have shown that, by means of chemical, optical, auditory, tactile and other electrical signals, there is communication not only among organisms of the same species and among organisms of different species but also among cells of the same organism and between parts of the same cell. Without inter- and intracellular communication, life would be impossible. All living things communicate and, according to the late Norbert Wiener, machines can communicate. Computers have proved this. The transfer of meaning is always involved. The word "communication" can be traced to the Latin word *communis,* which means commonness. When people communicate with one another, they establish a commonness; they share a commonality. Dictionaries define the process as "the giving and receiving of information signals or messages by talk, writing, gestures, and signals."

Sociologists tell us that the communication process is the basis of all social existence. It is basic to the development of the individual, to the formation and healthy existence of groups, and to the functioning interrelations among groups, organizations, cities, and nations. Communication links person to person, every person to the group, and the group to a larger encompassing social structure.

To understand the dynamic, living, and complex process of communication, we must break it down and examine its ingredients. If we base our analysis on the definition of communication, we shall come up with the following components.

1. The *source* or *sender* of the message.
2. The *message* being sent.
3. The method of the sending or *channel.*
4. The obstacles in the way—interferences or *noise.*
5. The *receiver* or *destination* of the message.
6. The *effect* produced in the receiver by the message.
7. The *feedback* or *reaction* by the sender to the effect of the message in the receiver.
8. The *frame of reference* or *environmental factors* in which the process is taking place and the influence the environment may have on the sender, the message, the channel, the noise, and on the receiver.

With eight ingredients or variables involved in a mix or interaction, the process is so complicated that it seems a wonder that communication is ever successfully consummated. But communication does take place.

HOW COMMUNICATION WORKS

What happens when communication takes place? First, the source receives a stimulus by means of his perception apparatus. Reacting to the stimulus, he formulates his message; that is, he takes the information or thought or idea he wants to share—taking it from a stockpile of possible messages—and encodes or expresses it in a form that can be transmitted. The coding process is the thinking that takes place in the mind of the source. The message is the thought, coded in a language and format that can be transmitted—for instance, a letter, sound waves in the form of spoken words, a picture, or a diagram. The message is received at its destination through the receiver's perception system. The receiver can understand the message only within the framework of his own stockpile of experience and knowledge. The source can formulate messages and the receiver at the destination can decode them only within the experience each has had. If there is no commonality of experience or empathy, then communication is difficult if not impossible.

If the person sending the message does not have adequate or clear information, the message is not formulated adequately, accurately, or effectively. If the message is not transmitted accurately enough (despite interference and competition) to the desired receiver, if the message is not interpreted in a manner that corresponds faithfully to the formulation, or if the person receiving the message is unable to respond or react to the message so as to produce the desired response, then there is a breakdown in the communication system.

The observation by the sender of the effect of his message on the receiver is called *feedback*. Feedback is an important means for gauging the success of a message. It can be as simple as the question, "Do you understand what I mean?" But a reply of "Yes, I do" does not necessarily assure that the message has been clearly received and assimilated. Further interrogation and observation of total reaction may be necessary to obtain assurance.

In a communication situation, each person is both a sender and receiver. The communicator gets feedback of his own messages by listening to his own voice as he talks or by reading his own letter as he writes. Thus we are able to correct mispronunciations as we listen, or to catch our mistakes as we read what we have written, or to add information as we recognize an omission. Feedback enables us to catch weaknesses in communication links where they may occur—especially if noise is present.

Books, newspapers, letters, and other printed and written forms eliminate the aids to the transference of meaning provided by the direct, face-to-face human interaction afforded in facial expressions, gestures, and intonation. In the written mode, an extremely important factor is immediately absent—feedback. The absence of immediate feedback requires the writer not only to plan his message more carefully but also to be knowledgeable about his unseen receiver and the receiver's psychological, sociological, and motivational sets. Written communication employs a rhetoric of its own. It depends most on words and syntax to provide the nuances, the shadings, subtleties, and differentiations that carry the transference of meaning.

Notwithstanding these problems, written communication has been (and is) an effective means for social interaction. And despite the versatility of face-to-face communication, the written form has been the one responsible for man's progress from the days of cave dwelling to space flight. Because of writing, succeeding generations have not had to begin over again. Writing has preserved for us what has gone on before and has allowed and inspired us to progress further.

Effective written communication is difficult, partly because the writer must translate inner experience into an outer form and the receiver, in turn, must translate an outer form into an inner experience. What the communicator wants to impart is not experienced in the form of the symbols with which he is trying to transmit the experience. Similarly, the receiver cannot respond to a message without first encoding it (that is, transforming it into a more personalized form of experience).

MEANING

Meaning is fundamental to communication. When we communicate, we are trying to establish a commonness with someone—we are trying to share some experience, thought, idea, feeling, or information. Therefore, *when communication takes place, there has been a transference of meaning.* The science of transference of meaning is known as semantics. The process of meaning is so basic to communication that the terms are often interchanged. We say, "Tell me what you mean"; "Write clearly so the reader knows what you mean"; "His letter was meaningful to me"; "The letter I received from our field man says nothing to me"; and similar things. Meaning, however, is not transmittable. It is not a label tied around the neck of a word, sign, or symbol. It is more like the beauty of a face, as Colin Cherry has observed, which lies altogether in the eye of the beholder.[1]

Semanticists have called attention to some significant characteristics of the symbols (words) used in messages. Words are representational; they are substitutes for objects or occurrences in our experience. They take the place of things perceived in the past or of things that can be imagined. This capability is called "thinking of" or "referring to" what is not here. Our minds, through the use of symbols, for example, can manipulate a million dollars, ten thousand tons of ore, one neutrino, or tens of billions of light years without having these things physically present. This fact is the reason that semanticists caution us that words and language are not the nonverbal phenomena that they represent.

Words and symbols do influence. They are able to cause sensations within us. Some words simply inform; others can be manipulated to arouse emotion or persuade the hearer or reader to a course of action. The intention of the communicator as well as his skill in the choice and arrangement of his words are factors that influence the way the receiver decodes or is affected by the message. If the

---

[1] Cherry, Colin, *On Human Communication,* The Technology Press of the Massachusetts Institute of Technology and John Wiley & Sons, Inc., New York, 1957.

communicator simply offers information about events or things as they are, without intending to convey his own feelings about them or to induce set feelings about them to the receiver, he will use language that restricts itself to verifiable facts. On the other hand, if he wishes to convey a subjective view of the situation or if he wishes to influence or persuade the receiver toward a certain action, he will choose and perhaps slant his words to evoke an emotional response. Because words and symbols have acquired emotional associations, the communicator must be constantly alert to the fact that the structure of the language we use does not always correspond to the structure of the world. He must also be aware that words that reflect one outlook to a communicator may (because of psychological, sociological, political, or cultural differences) reflect an entirely different outlook to the receiver.

CORRESPONDENCE AS A MEDIUM

People have communicated with one another in some way since mankind has existed. Since the invention of writing, the epistle—the letter—has been among the most used instruments of human communication. Today, despite the telephone, television, Telstar and the computer, the letter remains an effective and much used means for effecting transference of meaning not only on the social level but in business, industry, and in pushing back scientific frontiers.

In order to obtain a better understanding of effective correspondence principles and practices, we have examined the communication process. A correspondent (sender) sends a message (a letter or memo) via the U. S. mail (channel) to a receiver (reader) with some purpose in mind (effect). The message will not achieve its purpose unless the sender and receiver have a common frame of reference (commonness). Feedback in correspondence is not direct. The action taken or not taken by the receiver provides the feedback. The receiver taking the action desired by the correspondent is the measure of success of the transference of meaning. In correspondence, as in other forms of communication, all the components of the communication have to operate.

With this analysis in the background, we now examine other fundamentals and principles of technical correspondence.

2
THE PSYCHOLOGY, PRINCIPLES, AND STYLISTIC PRACTICES OF TECHNICAL CORRESPONDENCE

In Chapter 1 we examined the process of communication. In this chapter we shall observe how the process underlies the principles of correspondence. Communication is sharing; it is interaction. It is a two-way process; it involves stimulation and response; and it is reciprocal and alternating. Through feedback, the response evoked by one message becomes a stimulus and a message in its own right. How this works is very evident in face-to-face communication; this holds equally true for correspondence: no letter or memo is ever sent without the sender wishing for a particular response. The receiver of the letter reacts to the message. Whether the letter is thrown into the wastepaper basket or whether the receiver takes the course of action desired by the sender often depends on the success the sender has had in establishing a commonality with the receiver—the success of the transference of his meaning. Communication takes place successfully when the effect produced by the message is that which is intended by the sender.

Before encoding a message, the sender must be sure of his intent—his purpose for sending the message. He must know what effect or response he wants to produce. Many letters end up in the wastepaper basket because the writer confuses the content—the words within his letter—with its intent, its purpose.

PSYCHOLOGY IN CORRESPONDENCE

The technical person has schooled himself to be scientifically objective and impersonal. His reports and papers are characterized by a calm, restrained tone and an absence of any attempt to arouse emotion. Reports and papers are reconstructions of scientific investigations; their main purpose is to be informative, factual, and often directly functional. The meaning that scientific reports and papers seek to transfer is the knowledge and insight the research has revealed to the experimenter. The emphasis is on the knowledge and insight (the message), not on the reader (the receiver). Personal feelings are excluded; attention is concentrated on facts. The report or technical paper is intended for a wide range of readers (receivers); the letter is intended usually for a single, specific receiver. Neither the report nor the letter is written for the pleasure of the writer or receiver; both are intended for a practical objective. However, the distinguishing difference is the definite intrusion of personal elements in the letter. The underlying psychology of practice in modern business letters is that they are reader-centered. The approach is based on the premise that a business letter communicates an attitude as well as a message. The success of transference of meaning depends on the establishment of a personal relationship between the writer and reader. The emphasis in the letter is on rapport; in the report, on facts.

The letter may be defined as a message in writing addressed to someone for a specific purpose. An effective letter is one that accomplishes its purpose—one that achieves the intended result. Good results will not be obtained unless the writer can reach his reader—that is, establish a commonality with him. To do this, the writer must adjust or adapt his message to the reader. This is one of the most important principles in the psychology of correspondence. It is often referred to as the "You Psychology." The maxim is: put yourself in your reader's shoes.

The practical application of the principle involves two considerations:

1. The burden of adjustment or adaptation is on the writer.

2. To make an effective adaptation, the writer must know or visualize the reader.

LETTERS ARE PERSONAL

When a reader receives a letter or memo he wants to know how the message affects or benefits him. So when you dictate a letter, remember you are writing to a specific reader who is a human being. He will not buy your product or service merely because you want his business. He will not hire you merely because you want to work for him. He will not accept delay in the delivery of your product only because you are having procurement problems. He will not buy for cash just because you ask him to. If you wish your reader to respond favorably, your letter must be constructed in terms of benefit to him. You must convince him that, by hiring you, he will get financial returns. You must show him that the material you want to use in the product he ordered is worth his waiting. You must prove that his buying for cash is the best thing for him. Visualize your reader; tailor and personalize your letter specifically for him. Begin your letter with something that will be of interest to him. For example, if you were a customer and received a letter that began:

> Even though our records show we mailed you a copy of our catalog GN-53A which contained full ordering instructions, we note that you did not include catalog item numbers in your purchase order as called for in our catalog GN-53A. To simplify our ordering and shipping problems, we require orders to include our catalog item numbers. We are holding your order in abeyance pending receipt of this information.

The letter is careless, thoughtless, and offensive. Wouldn't it be better if the letter had been worded as follows?

> Thanks for your order. We can supply the insulation you want in 1/2", 3/4", or 7/8" thicknesses. Please let us know which thickness you would like. I am sending along a copy of our catalog GN-53A. Won't you turn to page 3 for item descriptions and catalog number identification of the thicknesses of the insulation you need. We'll go right ahead with your order as soon as we hear from you.

The second example (above), demonstrates the *you* attitude of modern correspondence practice. The *you* approach is not a phony psychology. It takes advantage of knowledge of human nature. Every one of us wants to be treated with consideration, good will, courtesy, and tact. The expression, "More flies are caught with honey than vinegar," has arisen from human experience. The productive individual understands human relations. He makes use of the principles of human warmth, human values, and courtesy in his relationships. The scientist who is determinedly, coldly objective as he examines the facts and ingredients bearing on his scientific work must, in his communication with others, interact as a human being. The *you* attitude does not mean that the writer must be saccharine, wishy-washy, and without conviction. If the situation calls for firmness, the writer should be firm—firm, but not rude or discourteous. Examine the following two examples.

> We were shocked to receive your complaint about the amount of our bill for renovating your new office space. This is the first time in our history that our services have failed to please a customer. We are proud of the discriminating clientele we serve, and who take pride in the appearance of their office premises. When you asked us for a bid for painting your office walls we told you the exact price for each wall and for the superior grade of paint we would use. You asked us to scrape your floors and repair the windows, doors and molding. You mentioned no price limitation for the rest of the work. That is why we assumed you had confidence in our established 25 years' experience in this city. Now we see that our beautiful job of complete renovation is not appreciated, even though you gave us the impression you wanted the handsomest offices in the area. We

should have protected ourselves with a specific contract amount. We'll know better next time.

Compare the above petulant and irritating tone with the courteous but firm approach below.

Thank you for writing us so frankly about the way you feel about our bill for renovating your new office space. The charges for painting the walls, we understand, are acceptable because they are based on an agreed amount. The matter in question is the remainder of the bill covering scraping of floors, and repair of windows, doors, and molding. Our charges for these services and materials were itemized. They are based on prices that have been in effect for more than a year and that net us only a fair profit. Your instructions to us were to renovate the office space completely and to repair any and all broken items in the offices. We conscientiously did just that; it was our desire to provide complete, competent and satisfactory service. These considerations may have led us to do perhaps a little more than you may have expected, but in view of the results accomplished, we believe our bill is not unreasonable.

KNOW YOUR READER

Correspondence (whether it is social, business, or technical) is human communication. A letter or memo is a message from one person to another person. As a mechanism of human interaction the letter or memo deals with human beings; underlying its operational process are the same concepts of human psychology that underlie all personal interaction. Although your letter may deal with a matter your reader wishes to know about, you cannot assume a ready-made interest. If you do not gain the reader's attention immediately, he will skim, skip, or stop reading altogether.

Effective writers learn to capitalize on the built-in interest all readers have in the mail they receive, and good writers are able to hold the interest of their readers from the start, to the middle, and to the finish of the message. They succeed in doing this by their understanding of their readers. Because most of us have never met (nor, perhaps, will ever meet) all the recipients of the business letters we send, we must not allow the lack of personal acquaintanceship to cause us to be ignorant of our readers. This lack of familiarity is more reason for careful analysis and study of the factors surrounding a reader's identity, character, and the psychological and environmental background. The more we know of his likes and dislikes, his wants, his ways, his hopes, and his fears, the more likely we are to establish a common linkage. The understanding of common problems and interests can serve as a common frame of reference for effecting communication.

When a writing task stares us in the face, we often become more concerned with what to say than with the effect our ideas and words may have. The value of an idea depends on the degree of success it will have in influencing the reader to take a desired action. The effective writer is always aware that his message has a specific purpose to accomplish. He begins with the purpose and finds ideas and words that will influence the reader toward his objective. If the message is to be understood and accepted, it must be tailored to the reader's intellectual capacity and must correspond with the level of practice in his technical and professional sphere. If both you and your reader are technical specialists, you can write in the language common to your specialty. If you do not use this language, the reader may feel that you are talking down to him. If you are not sure of your reader's knowledge of the specialized content of your message, use nontechnical terms; define or explain them carefully. However, do not use technical language just to impress the reader. Language is a tool of thought. Complex or specialized ideas often require specialized terminology for expression. Words must be precisely chosen and sentences carefully and grammatically constructed to convey the message.

STYLE

Conventions and folklore of business and technical correspondence, based on years of mispractice, have a way of lingering on and on. Some of these conventions die slowly, and some remain to deaden or irritate the reader. We seldom see in letters the vapid, starchy prose of the "elegant eighties":

> Most Esteemed Sir:
> Your favor of the 3rd inst. received and duly noted.

This kind of writing rests where it belongs—in museum archives—but many of its insipid offsprings still are clearly recognizable in these frequently found phrasings:

> With reference to your letter of September 3rd, you are hereby advised that . . . .
> In reply to your letter of January 4, we wish to advise that . . . etc., etc.

In the past fifteen to twenty years there has been an ever-increasing reaction, opposite to the pompous style of yesteryear—the forced friendly and preciously precocious approach. A natural, direct letter is inviting to the eye. But most readers are overwhelmed and embarrassed by the directness of the forced cheer and salesmanlike bluster of the following example, representative of this approach:

SECURITY PAPER COMPANY
Tonawanda, New York

Mr. Robert Chandler
Parker and Bellum Advertising
202 Park Avenue
New York, N. Y.

SUBJECT: *American Precision Laboratory Instrument Corporation Annual Report*

We can do all right by you, Bob—in the way of an impressive collection of samples of—annual reports printed on dull coated papers—for your presentation to this customer.

The material is on the way to you via the usual channels.

All of the items included in the portfolio are properly identified as to grades of paper represented. We like to make it easy for you in this way. In particular, of course, the samples demonstrate the distinctive printing results on Brilliant Ivory grade of paper as you especially requested of us. Also, many of the reports are issued by prominent advertisers. No other paper manufacturer in the world can provide such an effective import of end results, as the material so ably demonstrates.

So pour forth your charm, Bob, on the higher echelon people of American Precision Laboratory Instrument Corporation. We'll hold out a special souvenir for the Chandler family as bait if you succeed in hooking the big fish.

Now isn't this a challenge worthy of the supreme effort?

With heartiest of greetings to you from all your Security Paper friends, we are, as always,

                              Promotionally optimistic,

                              BENJY CLARK
                              Advertising Department

Benjy Clark's letter, despite its objectionable style and overbundance of cliches, has some virtue—its vitality and directness. Most readers, however, even among sales and advertising people, would find its overstatements and breezy pace unappealing and overly aggressive.

What is appropriate style for technical correspondence? First, let us be clear that technical English is not a substandard form of expression. Technical correspondence meets the conventional standards of grammar, syntax, and punctuation. Technical style may have its peculiarities—in vocabulary and, at times, in mechanics—but it is, nevertheless, capable of effective expression and graceful use of language. Style is the way a person puts words together into sentences, arranges sentences into paragraphs, and groups paragraphs to make a piece of writing express his thoughts clearly. Technical style in correspondence is the way you write when you deal with a technical or scientific subject.

Style varies with the writer and subject matter. It is a writer's arrangement of words through which his individuality is expressed as he communicates his ideas and purpose. The most effective style is that which most accurately encodes the thought content of the message into transmittible language. In technical correspondence, style has certain basic characteristics. The most important characteristics are clarity, precision, conciseness, directness, objectivity, unity, and courtesy.

CLARITY

Clarity in writing is the quality of being unambiguous and easily understood. The secret of clear writing is clear thinking. Some letters are difficult to understand because the writer did not fully understand what he wanted to say. His ideas and purpose were vague in his mind. Sloppy, illogical, or incomplete thinking causes much of the lack of clarity in technical correspondence.

PRECISION

Precision and clarity are often interdependent. Clarity is achieved when the writer has communicated his meaning to the reader. Precision takes place when the writer attains exact correspondence between the matter to be communicated and its written expression. Words are the symbols of ideas and the ingredients of the thought units called sentences. An effective sentence cannot be made from imprecise, incorrect, or inappropriate words. To use the proper words in the proper places—as style is sometimes defined—the writer must select the precise shades of meaning among synonyms at his disposal. If he means instrument, he will not use the word device, mechanism, apparatus, implement, tool, contrivance, or instrumentation. The word "law" has many synonyms: regulation, canon, statute, ordinance, rule, practice, custom; none can be used interchangeably. When a substance is plastic, does the word mean it is pliable, ductile, malleable, adaptable, pliant, deformed, formative, capable of metabolic transformation, or does it refer to a polymer resin, or to neoprene? Sloppy usage or errors in usage result in lack of clarity and in creating an unfavorable impression on the reader.

Faults in clarity and precision may result from the following shortcomings.

1. The writer does not know his subject matter well enough to write about it.
2. The writer, although generally familiar with his subject, is unable to distinguish the unimportant from the important. (The essence of the matter has escaped him.)

3. The writer knows his subject matter, but is deficient in writing or communication techniques.

4. The writer does not know his reader well enough to direct his message to the reader's level of understanding.

CONCISENESS AND DIRECTNESS

Concise writing saves the reader time and energy because meaning is transferred in the fewest possible words. It is brevity without sacrifice of completeness. Directness also increases readability; it eliminates awkward inversions and circumlocutions—roundabout expressions involving unnecessary words or details. By eliminating unnecessary details the writer sharpens the impression of the main point of the message. Consider this example:

> We have an independent distributing organization that keeps our engineers informed of customer reaction and complaints relating to the product. The design changes that are indicated as required by these reports and by others that are sent us by our service department are generally finalized by the original design engineer for that product.

Translated into concise and direct English, the revision flows with clarity and interest:

> If feedback from our distributors and service department shows that changes are warranted, the original design engineer makes them.

Or consider the following example:

> Our laboratory furnace effects complete combustion of various materials without offensive smoke and odors to irritate workers in the contiguous area.

The following sentence says the same thing more concisely, directly, and clearly:

> Our laboratory furnace burns wastes completely. There is no smoke or smell to irritate workers in adjoining areas.

Below are some cliche phrases that impede the clarity and flow of the message.

Cliches and verbose expressions	*Concise and direct*
avail yourself	use
care must be exercised; caution must be observed	be careful, don't
from the viewpoint of	for
in order to	to
in the event that	if
in the nature of	like

Cliches and verbose expressions	Concise and direct
in view of the fact that	because, since
is provided with, is equipped with	has, contains
it is imperative that	be sure that
large number of	many
hold in abeyance	wait, postpone
presently	now
prior to	before
taken into consideration	considered
be cognizant of	know, notice
it is indicated that	we believe
perform the measurement	measure
functions to transmit	transmits
hard and fast	strict
one integral piece	one piece
mobile vehicles	vehicles
make the necessary adjustments	adjust
prove of interest to you	interest you
the shipment will go forward	we will ship
in due course	soon, promptly, at once
at this writing	now
it is desired that we receive	we want, we'd appreciate

COURTESY, SINCERITY

Courtesy is a commandment that may not be broken in correspondence. Courtesy has been defined as the quality in a letter (or memo) that shows a sincere consideration for the reader. Discourtesy in letters results from the same reasons as discourtesy in direct human relations—carelessness, insensitivity to the feelings of others, curtness or impatience, inexperience, stupidity, and anger. Courtesy should not be confused with effusive "soft soap" or flattery. Courtesy is genuine sincerity; it often is an expression of tact in a situation charged with emotion. Some words or phrases inadvertently raise a reader's blood pressure:

You claim
You neglected to include
You are mistaken
We cannot understand why you
We were surprised to receive your letter
You should have realized
We can hardly believe
Your complaint
You say in your letter
We cannot grant your request
We demand

Sarcasm is an extreme form of discourtesy. The reader interprets the offending word or sentence as contempt of or an attempt at superiority. In face-to-face interaction, sarcasm can be allayed by a friendly smile. On paper, sarcasm is unfriendly and shows disrespect. It is a costly price for an inept form of wit, and is effective only in losing the good will of the reader.

UNITY, COHERENCE

Unity is the principle of oneness. Unity in a message—a letter, a memo, any communication—means that the components, as well as the whole, deal with one main idea, thought, or thesis. Unity is achieved in a sentence, paragraph, and letter when a single thought is developed, and related ideas are subordinated. A sentence is often defined as the expression of a single thought. If the thought is incomplete or if several unrelated thoughts are mixed together into the sentence, it obviously lacks unity. The item that follows is not a complete thought:

> Having traced the signal on the oscilloscope.

But

> The engineer traced the signal on the oscilloscope.

or

> Having traced the signal on the oscilloscope, the engineer soon determined the trouble.

are complete and unified sentences. The sentence below unsuccessfully intermixes several thoughts. The intended message is not clear. The sentence lacks unity:

> Included in our offer are a year's guarantee, a detailed manual of instructions, one week's free trial use of the equipment and you are entitled to our customary cash discount of 2% if you pay your bill within ten days from the date of invoice.

The sentence contains at least two distinct ideas. Although two sentences might be used to express them, a single coherent sentence might be composed:

> Our sale offers you these very attractive features:
> 1. A year's guarantee
> 2. A detailed manual of instructions
> 3. One week's free trial of the equipment
> 4. A 2% discount for cash within 10 days of receipt of invoice

The revision has unity; the minor but related thoughts are subordinated to the main thought. Coherence is one of the attributes of unity. To cohere means to stick together. A sentence is coherent when all its parts are clearly and logically connected with one another.

A paragraph, like a sentence, requires careful organization of its component parts. A paragraph is built around a single topic. The topic or central idea of the paragraph is usually expressed in a *topic* sentence. If the central idea is not expressed in a topic sentence, it is clearly implied by the organization of the other sentences in the paragraph. This method is effective only when the implication is obvious. In correspondence, it is advisable to have topic sentences within all paragraphs because a busy reader may not see the implication you feel is obvious. Moreover, the topic sentence serves as a guide not only to the reader but to the writer. Like an efficient traffic cop, it excludes material that does not belong within the paragraph. It controls unity, coherence, emphasis, and proportion of the paragraph. The topic sentence is usually placed at the beginning of the paragraph, but not always. It may appear at the end, in the middle, or in any position in the paragraph.

Irrelevant ideas in a paragraph often appear in correspondence at the cost of clarity, conciseness, and unity. The inexperienced writer will find that beginning his paragraphs with a topic sentence will help in the organization of his ideas. This is a useful technique to start the flow of necessary information on paper. Test your paragraphs by these criteria:

1. What is the central idea?
2. What must I tell my reader to support it or explain it?
3. Is there anything in it not related to that idea?
4. Are the sentences organized in a sequence that is sufficiently logical to support or explain the topic sentence clearly?

THE PROBLEM OF OBJECTIVITY

Style in scientific writing is characterized by objectivity. By long tradition, personal feelings are excluded; attention is concentrated on facts. The conventional belief is that the exclusion of personal elements and personal pronouns produces a style consistent with objectivity and that the use of the third person and of the passive voice places emphasis on the subject matter. As a result, much of technical and scientific writing is dry and lifeless.

This kind of approach is contradictory to modern conventions of business correspondence and to principles of communication. When lifeless neuter subjects have things occur to them, the reader is disinclined to become involved. Facts, unless produced by warmblooded people, are distant and cold. The puerile practice of over-objectivity is probably the reason why scientific journals have become archival and professional scientific conferences have become so popular, even though oral communication is inherently less efficient than written communication. Conferences provide opportunity for live people to talk directly to other live people. The audience is usually permitted to talk back in the form of questions to the speaker. Interaction and feedback permit an exchange and testing of experience. There is an interplay of passion, concern, heart. The sender and the receiver benefit. There is no reason why informative, functional writing cannot be objective and still include personal pronouns. Periodicals like *Physics Today, Science, Science & Technology,* and *Industrial Research,* for example, allow their authors to play an active role in the narration of their articles and research papers. Readers become involved in the activity related. There is transference of meaning; there is communication.

Technical correspondence cannot live and prosper within stultifying conventions of the archival scientific journals. Technical correspondence is a channel in which live, warmblooded human beings share experience, exchange thoughts and ideas, and persuade one another to take or not take a course of action. Within human interaction, the principles examined in Chapter 1 operate. Technical correspondence follows the conventions and style of business correspondence. Although business correspondence is usually more formal than social correspondence, the writer and reader strive for a tone of personal and harmonious relationship. There are frequent instances where warm bonds and loyalties are established and kept over many years without the correspondents ever meeting face to face. Rapport, courtesy, and sensitivity to feelings do not preclude integrity, do not obviate intellectual honesty, and do not prevent objectivity. The qualities of clarity, directness, precision, and coherence in technical style have an emotional effect by helping to create an appropriate intellectual impression.

Now let us deal with the planning of the letter.

3
PLANNING THE LETTER

The novelist, the playwright, and the poet plot their writing. They do not set things down haphazardly, casually. They give much time and thought to planning the scenes of interrelated actions, incidents, or emotions which, through their arranged interplay, lead the reader or viewer to the point of the story, play, or poem. The plot is a series of occurrences—components moving from a beginning through a logically related sequence to a logical and natural outcome. There must be a beginning, a middle, and an end. The plotting simplifies the telling for both the writer and the reader. It is a selective series of incidents which brings the reader through an interesting experience from the beginning to the denouement—the outcome. Plot focuses the reader's attention on the theme and develops it.

The process of plotting also is an essential tool of the scientist. In his terminology, the word is *planning*. The scientific methods are his conventions for organizing and conducting his research. In reconstructing the research in the form of a report or paper, the scientist follows an intellectual procedure analogous to that of the creative writer. The investigator, having begun with a problem, has accumulated data for answering it. Just as skillfully as the playwright or novelist, he examines, analyzes, and interprets the raw data (the raw material of life in the case of the creative writer). The scientist, after his analysis and interpretation, develops the answer to his problem, as revealed by his investigation. He frames the answer into a clear and concise statement. This statement is the thesis of his research. It is what he will want the reader of his report or paper to know about his research. The thesis sentence helps him outline his paper. It guides him in the selection of the experimental "incidents" that he must include in his writing to lead the reader to the conclusion he arrived at.

Effective letters are creative and result from similar thorough, painstaking planning. Planning pays off because letters and memos represent sizable investments. They are expensive not alone from the point of view of the writer's and reader's time involved in the composition and deciphering but from the point of view of the benefits or penalties associated with the response of the written message. Good will, sales, employment and professional or business opportunities gained or lost stem from the effectiveness or ineffectiveness of the transmitted message.

Too frequently, correspondents begin to write or dictate their letters without adequate consideration of the problem or situation to be met. Having received a letter requiring a reply, some writers start off with a burst of energy. A first paragraph expressing immediate reactions is dictated or written. Then the writer is stuck. Better sense tells him he has not presented his view to his reader as effectively as he might. Or, having gotten his feelings off his chest, he must now find a way to resolve the situation. He dictates a few more unsatisfactory sentences. He stops to think over what he has written and lamely adds another thought or two. The result is a disjointed, confused message, which is unintelligible to the reader. The reader disregards it or misunderstands it. The action or inaction taken starts a series of expensive investments of time, money, frayed relations, and an exchange of further letters which, unhappily, may never remedy the situation. No scientist or engineer would approach a serious technical problem without study, thought, and planning. Yet many of them approach problematic communication situations in an offhand manner, without premeditated thought prior to coming to grips with the situation in the writing.

Many correspondence situations are relatively simple. They occur so regularly and with so few differences in detail that they are considered routine. In this category, for example, are requests for catalogs or other published information; transmittal of the material called for in these routine requests; acknowledgments of orders; and followup to acknowledgments. However, in the seemingly routine situation, opportunities are often neglected. Customers and potential customers respond in kind to offhand treatment. Consider the following letter.

> Thank you for asking for our booklet on our exclusive extrusion process. I am pleased to send it on to you. If you have any questions or if we can be of service please call on us.
>
> > Sincerely yours,
> >
> > T. A. DEMARETT
> > Sales Manager

Compare the above letter with the next one.

> Thanks for your inquiry expressing interest in the Burgess Steel *Coolflo Process*. In addition to the information contained in our *Product Design Guide*, attached to this letter, we would like you to know we can offer additional services to help you to adapt this process to your products.
>
> The Burgess Steel *Coolflo Process* is completely and exclusively different from other extrusion processes. Finished parts come from the presses with smoothness, hardness, strength and precision, *without requiring any*

additional machining. Furthermore, all of these features are acquired from the use of low carbon, *low cost* steel.

One of our experienced application engineers would be glad to call on you at your convenience to answer questions or discuss possible applications of this process to your products. If there is such an application, we can help you save costs and achieve the advantageous features in your products made possible by this process.

We look forward to hearing from you.

Sincerely yours,
T. A. DEMARETT
Sales Manager

The first letter is not a bad letter. It handles a routine situation adequately. However, it suffers in comparison with the second letter. The second letter has received thought and has benefited from planning. The reader receives an impression that his query has been given individual attention and that the respondent is interested in his problems, has thought about them, and can offer knowledgeable assistance. The two letters illustrate that, even though many correspondence situations are composed of familiar elements and a routine solution may be easily arrived at, adequate and careful planning may reveal new data and provide a more complete, more creative, and more effective approach. Thus, adequate and careful planning will promote the response in the reader that the writer wishes to evoke.

I do not mean that short letters are ineffective. Many simple situations require simple, brief statements. A short letter can be complete, friendly, courteous, and effective. Consider, for example, the single, simple sentence in the following request.

Gentlemen:

Please send me a copy of your current catalog.

Sincerely yours,
A. W. STONE

The next exchange of short letters handles the requirements competently and courteously.

Gentlemen:

Our order No. P-3783 called for seven cartons of Herbicide CRG-3. Your shipment contained only five boxes. Five boxes were all that were indicated in the shipping papers.

Please check this matter for us. We urgently need the two additional boxes.

Sincerely yours,
THOMAS S. BENSON

Dear Mr. Benson:

I am sorry you didn't receive all of the CRG-3 Herbicide you ordered.

We thought our shipment was complete, but you are right—two cartons are still due. They are being shipped today.

Sincerely yours,

OLIVER CREWPENNY

Letters and memos are messages written for a specific purpose. The message is composed clearly and succinctly in such a way that the reader receives the thesis or idea intact. To accomplish this, the letter must be carefully planned. The real work of composition takes place before the dictation or writing begins. No two correspondence situations are exactly alike. The experienced dictator may be able to handle familiar situations in a routine way. His planning becomes a matter of organizing and classifying matters from past experience in order to meet the new situation. Simple letter problems often can be met by establishing routines. But even the experienced correspondent who is dealing with an unusual or complicated situation examines closely, analyzes and plans carefully before writing, and frequently rewrites and revises.

The purposeful composition of a letter or memo involves five distinct steps:

1. Determining your purpose.
2. Considering your reader.
3. Gathering the information.
4. Outlining and organizing the letter or memo.
5. Writing the letter or memo.

DETERMINING YOUR PURPOSE

Before starting to dictate or write, ask yourself, *What do I want this letter to do? What action do I want the reader to take? What impression do I want to leave with him?* This type of an analysis to evolve the purpose of the message has the advantage of enabling the writer to start quickly and to get to the point of the letter. If the dictator or writer knows what he has to say and why, he will state it clearly and definitely; if he is unsure of his purpose, he will make a lot of noise and create confusion.

If you have to answer a letter, reread your correspondent's letter; underline questions and statements to be answered; jot down comments on the margins. In composing a reply, it is often helpful to examine past correspondence. When you have gained all of the background and facts, ask yourself, *What is the most important fact or factor to the reader?* Usually, the most important factor should be dealt with first; let the rest follow in logical sequence.

CONSIDERING YOUR READER

You have been considering your reader in determining your purpose. Many letters and memos fail because the writer is self-centered, concerned only with the information he wishes to transmit; he lacks the *you* attitude. In considering your reader, you should adapt your message to his needs and to the limitations of his knowledge. An exchange between two engineers familiar with a process or instrumentation has a different tone and less detail than a letter from an engineer to a customer offering instructions or explanations on that process or instrumentation. Even when addressing a company or an organization, the writer must visualize the member of that organization who will be reading the letter and making the decision on its contents. The writer must picture the situation he is expressing from the viewpoint of the recipient. Only then can he mobilize the facts to persuade the reader to his view. You need to consider not only items of positive interest for the reader but also probable objections to be overcome.

GATHERING THE INFORMATION

In the previous steps, the parameters of the amount of information required entered the considerations that defined the purpose and the reader. The next step is to assemble the information that will compose the essentials of the letter. Some situations are relatively simple. The information is readily available in the files of the organization or in the minds of the writer, his colleagues, or his superior. After some thought, analysis, and assemblage, the facts and elements are arranged into a logic determined by the nature of the basic message, the purpose, and reader requirements. In complex message situations the information must be gathered by investigative procedures based on one or several methods:

1. Searching of files and publications.
2. Observation, examination, and analysis of the situation or the factors involved.
3. Interview and discussions with experts or persons qualified to provide the needed data.
4. Research and survey.
5. A combination of two or more of the above.

OUTLINING AND ORGANIZING

After you have gathered all the information you need, you are ready to organize your message. The outline is a schematic road map of the message and shows the order of thoughts or topics and their relationships. The outline makes the writing of your letter or memo easier and more effective, and it enables you to dictate or write with confidence. It permits you to *control* the content and structure of the message in order to secure unity, coherence, and emphasis.

Relatively simple situations require simple organizational schemes. The purpose of the letter or memo is placed at the top of the outline. The thesis or main point to be delivered to the reader is placed immediately below the purpose. Then the topics for accomplishing the purpose and establishing the thesis are set down. More experienced correspondents are able to plan the outline in their minds, but until that capability for organizational planning of letters and memos becomes second nature to you, you should record the outline on paper or on the margin of the letter you are answering. The outline should be simple in form and content. Set down topical headings for each major point or idea. The topics need not be recorded in complete sentences, but they should be specifically stated.

For example, the letter dealing with the two missing cartons of herbicide might be outlined as follows:

    Purpose: Inform Seller Shipment Incorrect
    Thesis: Urgent Need of Two Additional Cartons
       1. Order called for seven cartons
       2. Five cartons received
       3. Shipping papers said five cartons
       4. What happened?
       5. Check matter
       6. Urgently need two more cartons

The outline facilitates an orderly arrangement of ideas and statements so that the reader can follow the planned thought of the letter from beginning to end. Careful letter planning will show dividends as the writer begins dictating, penning, or typing the letter.

WRITING THE LETTER

Letter and memo situations vary with individual problems. A letter that proves effective for one situation will not be appropriate for another. Since recipients and circumstances differ, a set formula for letter construction is not practical.

If we examine the reading habits of letter recipients or consider our own letter reading, we will notice that the first and last paragraphs of a letter receive the most attention. They are given the most attention because they are read first. With so many pressures claiming the time of readers, letters receive quick readings or rapid scannings. If the opening paragraph catches the reader's attention, he will go on, often skimming the middle portion and looking to the final paragraph for the point or thesis of the letter, or for the action or decision he is called upon to make.

This awareness of letter-reader habits provides a lesson in letter writing. The opening and closing paragraphs are psychologically—because of the emphasis inherent in their position—the most important elements of your letter. The beginning paragraph introduces you and your subject (thesis) to your reader. This paragraph must favorably grasp the reader's attention. The final paragraph is the one he reads last and, therefore, is freshest in his mind. By psychological position, it should reinforce the purpose of your message, and it should motivate the reader to take the action you desire.

If the opening paragraph of your letter is fresh, direct, and interesting, the entire letter may be read. If the beginning sentence is trite and dull, the rest of your message may go unread or, at best, may be scanned. Check the carbons of the letters you have written. How many of them have stereotyped beginnings such as the following ones?

> Regarding your letter of June 18 in which you request . . . .
>
> We have your letter of April 14 in which you state that delivery of your order #P0243 has not been received as promised . . . .
>
> This is to acknowledge with many thanks your generous offer for . . . .
>
> I have before me your letter of November 9 in which you inquire about my knowledge of Mr. Stanley Hunter who worked for this Company from June 15, 1963 to July 1, 1968 . . . .

Do you recognize any of those opening words?

Those above beginnings are ineffective because they are shopworn, pedestrian, and weak. They have no human interest or warmth. They are composed of stereotyped words that dull the reader's senses. The opening paragraph of a letter has four duties to perform:

1. It must attract favorable attention.
2. It must have a bearing on the purpose and point of the letter.
3. It must set a tone of courtesy and friendliness.
4. It must, if appropriate, connect the letter to previous correspondence by a reference to date or subject.

Openings that have these four values tend to be more direct and interesting. Letters that start with a detailed summary, rehashing the letter being answered, deaden the reader's interest.

An effective opening starts with news that the reader is waiting to hear. It is much like the headline of a newspaper—it telescopes and promises information that is of interest to the reader. When reference to previous correspondence is necessary, the reference should be subordinated—not featured. Like the headline, the opening announces to the reader what the letter is all about.

But the opening paragraph should do more than tell the reader specifically what the letter is all about—it also should create a favorable impression upon him. The first impression affects the reader's attitude toward the rest of the letter. If this impression is unfavorable, the reader will be unreceptive—perhaps hostile to the letter's purpose. If the impression is neutral, the reader will be neutral and unlikely to take the action desired. If the opening creates a favorable impression—if it catches the interest of the reader and is courteous and gracious—it will induce the reader to be kindly disposed to the rest of the message and to the desires of the writer.

Slow and ineffective

We are referring to your inquiry of February 6 in which you requested information regarding our continuous-wave gas laser altimeter. . . .

Direct and effective:
(The identifying date information is made subordinate to the more important and interesting topic the reader is interested in.)

Here is the information you wanted on our continuous-wave gas laser altimeter. You may be interested to know that the Federal Aviation Agency has voted that airliners equipped with systems utilizing our Altimeter may land with as little as a 100-foot ceiling. . . .

Trite, involved, inappropriate and ineffective:

We acknowledge and thank you for your letter of December 18 in which you call our attention to the fact that you have not received your order No. 19872 for 17 magnetron electronic cooking ovens, our model No. 2P73. . . .

Direct and effective:
(There is no need to acknowledge and certainly the opening of thanks is inappropriate. The rewrite comes immediately to the point the reader wishes to know.)

Ten of the 17 magnetron electronic cooking ovens, model No. 2P73, have been shipped to you today, rail express. Six of the remaining seven will be shipped two weeks from today or perhaps one day sooner, and the seventh bearing the modification you requested in your change order will be ready for shipment and sent six weeks from today.

The opening and closing paragraphs of your letter are critical, but the paragraphs in between carry the load of the message. They should flow directly and explain to the reader the *who, what, when, where,* and *why* of the message.

You should determine the number of paragraphs—one for each major thought or fact—before you start writing or dictating. This planning will force you to order your thoughts. Make use of 1-2-3 or a-b-c lists wherever possible. This type of organization promotes understanding and provides emphasis.

Try to visualize the closing paragraph before you begin; keep moving toward it when you dictate or set down your beginning sentences. Construct the last paragraph to make it easier for your reader to take the action you want him to take. Or use this paragraph to build a favorable attitude when no action on his part is needed. When your message is completed, stop.

Too often, inexperienced writers feel uneasy about the end of their message. They have said all there is to say but, like the proverbial dinner guest who does not know its time to stop talking and depart, go on, and on, and on. Often the inexperienced writer will tack on a participial phrase or an indefinite, stereotyped thought that hangs on foolishly like a jarring anticlimax:

a. Trusting we can be of service, we are . . .
b. Thanking you in advance, I am . . .
c. We thank you for the opportunity to be of some service and sincerely hope the above information meets your approval.
d. We hope you can take care of this matter . . .
e. We shall appreciate your kind attention to this matter at your earliest convenience if possible.
f. We wish to thank you for your interest in our products and hope you know we are waiting to serve you as you may require our services.

More effective closings for the above examples are:

a. If you have a special measurement problem, our engineers might be able to help you. Just call us, and we'll be glad to send someone to see you.
b. Can you let us know by October 1 whether you will have the high speed digital printer for delivery.
c. Please send us the completed forms by May 15. We'll do the rest.
d. Since both our records agree as to no further purchases, we look forward to the explanation why our credit balance is lower this month than in January.
e. Won't you send us your check not later than March 15.
f. We are glad to send you the brochure describing our avionic products. On page 16 you will find the answers to your specific questions on design and special applications. Mr. Tom Egan, an engineer in our Federal Systems Division, will be happy to provide any further information on any of your special needs. His telephone number is 496-8888. Please call him. He will be glad to help.

If you follow the five steps in planning and writing your letters and memos, your messages will be more effective as communication instruments because of the logic of their organization, their directness of approach, their sincerity and character in expression, and because of their warm, friendly, and courteous attitude. The readers of our letters are humans—just as we are. We must reach them through the warmth of human qualities and the logic of human intelligence.

4
MECHANICS OF CORRESPONDENCE

The message of the letter is more important than its format or style; however, we should recognize that style and format are to a letter what dress and appearance are to a man. None of us would appear for a job interview in a sweatshirt, unkempt and unshaven. Similarly, when we send a letter, it speaks for us in the correspondence situation. A single error may nullify an otherwise well-written letter. The receiver of the letter can evaluate the message only by the letter's total impression. If even a minor aspect suggests carelessness, slovenliness, or inaccuracy, the reader loses confidence in the more fundamental worth of the message. Which of us would want to be operated on by a surgeon who spells appendicitis the way the word is pronounced? An examination of the format, style, and conventions of business correspondence is important because readers begin reacting to the immediate impression of a letter or memo even before the significance of the content has filtered through. Instantaneous, first impressions color later reactions; therefore, outward and stenographic details of correspondence practice are important considerations.

STATIONERY CONSIDERATIONS

Standard business stationery is 8½ by 11 inches. Half-size letter sheets (8 by 5 inches with the standard letterhead or 5½ by 8½ inches with a half-size reproduction of the letterhead) are sometimes used. United States government offices and military services have standardized 8 by 10½ inch letter stationery. Generally, in the technical situation (other than the military and government), it is better to use the 8½ by 11 inch letter sheet.

The first item that your reader will notice is your stationery. A good quality 20-pound white bond paper is always in good taste. Heavier paper is expensive; it is too stiff for easy folding and too thick for carbons. Paper lighter than 16-pound is too flimsy and too transparent for letters. Buff and off-white blue are sometimes used. Certain commercial organizations prefer a colored stationery because they believe their correspondence stands out from other correspondence in a file and attracts attention. Choice of color should be governed by the appropriateness of the color to the character of the organization. Colors that depart radically from white may encounter prejudices on the part of readers. A good quality water-marked, white, rag-content bond lends dignity and is in keeping with the formalities of the letter situation; it is the stationery most frequently used.

The design of an organization's letterhead should be left to specialists in printing and paper supply companies, who know paper stocks, color, and graphics, and who can manipulate these to best characterize the organization that the letterhead identifies. Considerations of this kind are beyond the scope of this book. Letterhead stationery is used when appropriate. It can serve as both a mechanical aid for more effective layout of the message on the page and as a psychological aid to draw favorable attention to the message by the role it plays in attracting the reader's eye and framing the message on the page.

FRAMING THE LETTER

A typed letter should be placed on the page so that the white margins serve as a frame around it. Practice and convention require more white space at the bottom than at the top unless the length of the letter and the size of the letterhead make this impossible. The side margins should be approximately equal and usually should be no wider than the bottom margin. Side margins conventionally are one to two inches, depending on the length of the letter. The bottom margin should never be less than one inch below the last typed line. The right-hand margin should be as even as possible. The body of the letter proper should be placed partly above and partly below the center of the page. A short letter makes a better appearance if it is more than half above the center of the page. The placement of the letter on the page (frequently referred to as "centering") depends, of course, on the style of layout used. (Conventions of layout styles will be explained shortly.) Centering is achieved by starting the first line of the inside address at a chosen depth to give the most pleasing appearance of white space above and below the message. The shorter the letter's message, the lower the inside address is placed.

STANDARD LAYOUT STYLES OF LETTERS

In the interest of efficiency, most companies and organizations encourage a standard layout style for all of their letters. The three most commonly used forms today are the *semiblock* (sometimes called *modified block*), the *block,* and the *full block* (sometimes called the *simplified style* or *NOMA* form). Examples of these conventional layout styles follow.

GEOSCIENCE APPARATUS CORPORATION
290 River Road
Leeds, Massachusetts 02127

September 13, 1968

Dr. Herman M. Weisman
7807 Hamilton Spring Road
Bethesda, Maryland 20034

Dear Dr. Weisman:

    Thank you for the opportunity to help you prepare your students for what the technical community will require in letter writing. The importance of this professional skill cannot be emphasized too much, for the ability to write effective letters is a basic requirement for professional success.

    Our corporate training courses on writing stress three things:

1. Be brief. -- Professional men do not read letters for pleasure, so make them short and to the point.

2. Be specific. -- Do not make the reader interpret your letter.

3. Be conversational. -- Stilted, formal letters build a barrier between the writer and the reader.

    The style of letter format preferred by our company is the semi-block form as illustrated by this letter. We like this style because experience has told us many business men prefer it, since it represents a happy compromise between the old fashioned indented form and that of the radical modern full block.

    If you would like any additional information, please call on us again.

                                  Yours very truly,

                                  GEOSCIENCE APPARATUS CORPORATION

                                  M. G. McDonald, Manager
                                  Administrative Department

MGMcD:wj
Encl.

AMERICAN CYROGENICS COMPANY
6200 Holyoke Boulevard
Stamford, Connecticut 06904

October 1, 1968

Dr. Herman M. Weisman
7807 Hamilton Spring Road
Bethesda, Maryland 20034

Dear Dr. Weisman:

If our practices in business correspondence can be of help to you and your students, I am happy to share them with you. Examples of our approach are attached.

A year ago we issued a <u>Secretarial Guide to Style</u>. Until we issued this style manual we had a wide variance of set-ups for letters, differences in approach to punctuation, numerals and abbreviations. Consequently, our letters gave a wrong and, I am afraid at times, a weird impression of American Cryogenics Company.

A standardized and uniform style in our correspondence has not only made our letters and memos more readable, improved the appearance of our company image, but also increased secretarial efficiency.

Efficiency is the reason we have adopted the block form in our letter layouts.

We are glad to cooperate in your survey and hope our views are of some help.

Sincerely yours,

Frank Gardener

Frank Gardener
Head, Education Department

FG:ta
Encls.

DORRENCE ASSOCIATES
426 North Michigan Avenue
Chicago, Illinois 60611

17 July 1968

Dr. Herman M. Weisman
7807 Hamilton Spring Road
Bethesda, Maryland 20034

Dear Dr. Weisman:

We are public relations consultants. Building favorable public images is our business. Naturally we are concerned with our own. That is why we have chosen to use the simplified letter form as the correspondence format style to reflect our character.

We are modern, bright, efficient. We are on the go! Not the semiblock, nor block -- and <u>certainly not</u> the indented style would represent our personality.

The characteristics of the simplified style are illustrated in the layout of this letter. Not only do our secretaries find its specifications easy to follow, but also our customers and public have favorably received it.

Sincerely,

Sam K. Dorrence

Sam K. Dorrence
President

SKD:lp

THE RONALD MICRO CORPORATION
1048 Terminal Avenue
Bellaire, Texas 77401

October 1, 1968

Re: Micro Reader
    Refer to 201.00

THANK YOU, Mr. Abbey,

for your inquiry and interest in Ronald Micro Corporation pocket micro readers.

Accordingly, we are happy to send you the enclosed information covering this unique precision unit which meets the most exacting standards in the microfilm field. In addition, we are sending you information on other pocket precision viewers manufactured by us, which should be of interest to you.

Just add microfilm and read...anytime...anywhere... they're versatile, compact and inexpensive...test them... satisfaction guaranteed.

Please feel free to call on us should you require any additional information. We are at your service.

Cordially yours,

THE RONALD MICRO CORPORATION

Harvey B. Sanders
Harvey B. Sanders
Director of Sales

HBS:mg

MARLOWE'S MEN'S STORE
552 Madison Avenue
New York, New York 10016

August 11, 1968

Mr. Laurence C. Groves
618 W. 81st Street
New York, N.Y. 10025

Dear Mr. Groves:

Marlowe's Men's Store cordially invites you to see a preview of
our new fall fashions for men Thursday evening
August 14.

The showing will be on Marlowe's second floor at 8:00 PM. The
doors will open at 7:30. Tom Blick, football coach
of the New York Jets, will be on hand to welcome you.

In our 1968 show you will see every type of new fall men's wear
from informal golf attire to black tie and tails.

To commentate on the new styles in men's clothes will be Jerry
Lubet, the talented, bright new MC of the <u>After Midnight</u>
TV show on Channel 8.

Of course Tom Blick will talk football "between the halves" at
8:30. There will be a surprise door prize. So be sure
to join us, won't you? No tickets are necessary.

Cordially,

Nick Marlowe

Nick Marlowe
President

Some organizations use all three major styles or combinations of them. Many companies issue style manuals for letters and memoranda in order to achieve uniformity and excellence. If the organization or company for which you work has such a manual, follow its recommendations.

The semiblock format differs from the block format only in that the paragraphs in the body of the letter are indented. The block and semiblock formats are used in about 80 to 90 percent of all typed business letters. The full block or simplified letter is very efficient in that it eliminates the need of indenting and tabulating by the stenographer, but many readers are disturbed by its unbalanced appearance. Management-consultant firms have a tendency to use this form because it gives the appearance of efficiency. Many public relations firms and advertising firms also use it because of its breezy appearance. The simplified letter form was originated by the National Office Management Association. The "NOMA" letter is more radical in style and appearance than the simplified form. It omits the salutation and frequently uses instead an attention-getting device of a subject line in capitals. It often dispenses with the complimentary close and may omit the dictator's and typist's initials. An example of the NOMA letter is shown on page 38.

Two other styles (used with decreasing frequency) are the hanging indentation and indented styles. The indented style is rarely, if ever, used today. All of its elements—heading, inside address, and complimentary close—are indented. Because of its built-in inefficiency it has lost favor in today's fast-moving world. I mention it only as a historical curiosity. The hanging-indentation format is used at times by advertising firms, especially direct-mail concerns. The first line of each paragraph is flush left. Succeeding lines of the paragraphs are indented—hence the term hanging indentation. The effect emphasizes the first words of each paragraph. The unusual appearance of the letter attracts attention; this is, a psychological factor appropriately significant in bringing the message to the reader.

MECHANICAL DETAILS

Components of the Letter

Conventionally, a letter may have from six to ten components. Some components may not be required in some letter situations, as I shall explain later.

Heading

The preprinted letterhead forms the element of the *heading*. When preprinted letterhead paper is not used, the writer includes a heading to help identify the source of the letter. Then the heading element includes the address, but not the name of the writer. The street address is on one line, and the city, state, and zip code on another. The date follows the city, state, and zip code line. The heading is placed at the right side of the page in block form:

    3375 South Bannock St.   1230 65th Street
    Englewood, Colorado 80110   Milwaukee, Wisconsin 53214
    May 16, 1968   June 2, 1968

                      220 Columbus Avenue
                      Boston, Massachusetts 02116
                      February 28, 1968

The heading helps to frame the letter on the page. The usual upper margin is 1½ inches from the top of the paper. If the letter is short, the white space of the top margin will, of course, be deeper. The placement of the heading should be planned so that the line ends at the right-hand margin of the letter.

The Date Line

The purpose of the date line is to record the date the letter is written. When preprinted stationery is used, the date line is the first item typed. If ordinary bond paper is used, the date line has become the last element of the heading. (Notice the examples above.) The date line can be centered or placed on the right or left depending on the style of layout. Often the design of the letterhead will suggest one position or the other. The date line is frequently placed so that it ends flush with the right-hand margin of the letter. Its usual placement is about four spaces below the last line of the letterhead. If the letter is very short, the date line should be dropped to give a better balance to the page. The month should be spelled out in full and no period is used after the year.[1]

Inside Address

The purpose of the inside address is to identify the receiver of the letter by giving the complete name and address of the person *or organization* to whom the letter is being sent. The address is usually placed four to six lines below the date line. No punctuation is used at the end of any line in the inside address. No abbreviations should be used in the address. Write out street or avenue, as

---

[1] In the past, it was customary to punctuate the lines of the heading and inside address with commas at the end of each line of the components, except for the last line that received a period. In recent years, practice has been to omit all end punctuations except for periods at the end of abbreviations. Commas, however, are used between the names of the city and state and between the day of the month and the year.

well as the name of the city and state. Use a courtesy title, such as *Mr., Mrs., Miss,* or *Dr.,* when addressing an individual. When a title follows the name of an addressee, it is written on the same line except when it is unusually long; then the title is placed on the next line:

Mr. Scott Rattet
4870 North Fifth Avenue
Chambersburg, Pennsylvania   17201

Dr. Harlan F. Weisman
Director of Research
American Biological Laboratories
8000 Georgia Avenue
Silver Spring, Maryland   20910

Mr. John Scoefield, President
Giant Steel Corporation
9700 Rock Street
Rockford, Illinois   61101

Attention Line, Subject Line, Reference Line

The Attention Line is a conventional device used to direct the letter to a specific person, to the individual occupying a certain position, or to the department especially concerned. An Attention Line provides a less personal way of addressing an individual than by placing his name at the head of the inside address. It is a convention that seems to be losing favor, since modern practice is to address letters to individuals.

The Subject Line (popularized by military correspondence, the interoffice memo, and by the NOMA approach) is an efficient device because it helps the reader to know immediately the thesis of the message. Its use also avoids the stereotyped letter beginnings that refer to letters being answered.

The Reference Line also avoids beginnings that refer to previous correspondence or to files of documents, invoices, or similar matters.

The position of the Attention Line is usually between the inside address and the salutation. It is placed two spaces below the last line of the inside address; it is not capitalized, nor is it usually indented:

E. H. Mitchell & Co.
1834 West Foster Avenue
Chicago, Illinois   60630

Attention: Dr. S. A. Ross, Chief Engineer

If a Subject Line is to be used, it should be typed below or in the place of the Attention Line:

Rovere Scientific Instruments, Inc.
One Bridewell Place
Clifton, N. J.   07014

Attention: Mr. Foster Wells, Sales Manager
Subject: Photoelectric Colorimeter Model 10A

No end punctuation marks are used with the Attention, Subject, and Reference Lines. Practice varies on exactly where these components are placed. Since the purpose of these devices is to call attention to and identify immediately the person to whom the letter is directed, or the subject word matter to be discussed, they should be placed conspicuously at a point strategically at the beginning of the message. Because of the latitudes found in practice, I shall not be prescriptive but shall indicate the conventions in use. In some instances there seems to be little difference between Subject and Reference Lines. Some company correspondence manuals use one or the other. Sometimes the abbreviation of reference (RE) is used.

The Subject Line may be placed on the same line with the salutation, or two line spaces below the last line of the inside address. Here are some examples.

1. Carl Jenson Chemicals
   444 Fifth Avenue
   New York, N. Y.   10018

   Subject: *Solvents for Pesticide Residues*

2. Carl Jenson Chemicals
   444 Fifth Avenue
   New York, N. Y.   10018

   Gentlemen: *Solvents for Pesticide Residues*

3. Carl Jenson Chemicals
   444 Fifth Avenue
   New York, N. Y.   10018

   Solvents for Pesticide Residues

   Gentlemen:

4. Raytheon Company
   South Norwalk, Conn.   06956

   Reference: *Our Purchase Order No. P8961*

5. Raytheon Company
   South Norwalk, Conn.   06856

   Gentlemen: *Our Purchase Order No. P8961*

6. Raytheon Company
   South Norwalk, Conn.   06856

   Re: *Our Purchase Order No. P8961*

Some organizations use file numbers or classification numbers that make reference to or identify the subject or the writer of previous correspondence. File or classification numbers are usually positioned where the Subject or Reference Lines are ordinarily placed. For example:

44 Principles and Fundamentals

    Lee Ultrasonics, Inc.
    Waltham, Massachusetts 02154
    File No. 201.03

Government agencies, insurance companies, and some other organizations who have devised filing or classification systems frequently preprint a place for a file or reference number below the date line. For instance:

    U. S. Treasury Department
    Internal Revenue Service
    District Director
    P. O. Box 538
    Baltimore, Maryland 21203

                                          June 29, 1968
                              File No. 5221307203929-68

The first letters of important words in the Attention, Subject, and Reference Lines should be capitalized. Underscoring may or may not be used, depending on personal preference.

Salutation

    The salutation originated as a form of greeting. The NOMA and simplified letter forms omit it entirely as being old fashioned and superfluous (see letter example pages 37 and 38). Most readers are jarred or offended when the salutation is omitted. It is placed two line spaces below the inside address or six spaces if an Attention Line or a Subject Line is used. Use a colon after the salutation. Conventionally, the word *Dear* precedes the person addressed unless a company or group is the recipient of the letter:

    Dear Mr. Wilson: *or*
    Dear Sir:

    Dear Dr. Brady: *or*
    Dear Sir:

    Dear Mrs. Rice: *or*
    Dear Madam:

Gentlemen: is the correct form in addressing a company, organization, a group of men, a post office box, or a newspaper number. *Ladies:* or *Mesdames:* is correct when addressing two or more women; however, if a company bears the name of a woman, its proper salutation is *Gentlemen:*

    Helena Rubenstein
    500 Fifth Avenue
    New York, New York

    Gentlemen:

Abbreviations should not be used in the salutations other than *Mr., Mrs., Miss, Messrs.,* or *Dr.* Titles like *Professor, Colonel, Governor, Senator,* and *Secretary* should be written out. The first letter of titles in the salutation, including *Sir,* should be capitalized.

Body of the Letter

Begin the body of the letter two line spaces below the salutation. Whether the paragraphs of the body of the letter are indented depends, of course, upon the style used. Single-space the letters of average length or longer. Short letters of about five lines or less may be double-spaced. Always double-space between paragraphs in single-spaced letters. Quoted matter of three or more lines is indented at least five spaces from both side margins. Very short letters that are double-spaced should use the semiblock form to make paragraphs stand out as separate units. Double-spaced material does not receive extra space between paragraphs.

Complimentary Close

Just as we say goodbye when we leave a person, the complimentary close (by convention) expresses farewell at the end of the letter. (In the NOMA letter the complimentary close is omitted as superfluous.) The complimentary close is typed two lines below the last line of the message. It should start slightly to the right of the center of the page, but it should never extend beyond the right margin of the letter. The comma is used at its end. Only the first word of the complimentary close is capitalized. The following are the proper and conventional complimentary closes that are most frequently used.

>   Yours truly,
>   Very truly yours,
>   Sincerely,
>   Sincerely yours,

Cordially yours implies a special friendship; *Respectfully yours* implies that the person addressed is the writer's social or business superior.

Linking the body of the letter with the complimentary close by a participial phrase is improper, ineffective and, often, ungrammatical:

>   a. Thanking you for your interest we remain,
>
>   Very truly yours,
>
>   b. Hoping we have been of some help in your problem we are
>
>   Sincerely yours,

Correct and effective:

a. Thanks for your interest. Give us two weeks to study your suggestion; we will then get in touch with you.

Very truly yours,

b. We were glad for the opportunity to examine with you your requirements for surface charge leakage measurements. If you have any other electrostatic field measurement problems, please feel free to call on us again.

Sincerely yours,

Signature

When the sender's name and title are printed on the letterhead, the typing of the title is frequently omitted after the signature. Otherwise, the name and title are typed in block form three to five spaces directly below the complimentary close. The first letters of each word are capitalized; no end punctuation is used. The letter is neither complete nor official until it is signed. Only after the dictator or writer has signed the letter does he become responsible for its contents. Read your letter carefully before you sign it. Once it goes out, *you*, the signer, are responsible for typographical errors or other mistakes. In routine letters, a secretary may at times be delegated to sign the name of the official sender. In such a case, the secretary or stenographer signing the sender's name adds her initials in ink. Sometimes her initials are preceded by the word *per*.

Yours truly,	Sincerely yours,
Glenn C. Britton Director of Research	Kermit I. Rosen President
Respectfully,	Cordially,
Logan Boardman Systems Analyst	Sheldon B. Kornberg; Sc.D. Head, Electric Engineering Department

If a company instead of the writer is to be legally responsible for the letter, the company name should appear above the signature. The use of a company letterhead does not absolve the writer; therefore, if you want to protect yourself against legal involvement, type the company name in capitals, leaving a double-space below the complimentary close; then leave space for your signature before your typed name:

Yours truly,
UNION CHEMICAL CORPORATION

Sincerely yours,
DENVER BIONICS

Mark Slocum
Process Engineer

T. S. Elliot
Technical Writer

Identification Line

In order to identify the dictator or official sender and typist, the dictator's initials followed by those of the stenographer are typed two line spaces below the signature, flush with the left-hand margin. The dictator's or sender's initials are typed in capitals, with a colon or a slant (/) and the typist's initials follow. For example:

HMW:sr or HMW/sr

If the letter is written by a subordinate but signed by his superior, the identification line will show this. For instance:

OWR:HMW:jh

Enclosures

If your letter contains enclosures, type *Encl.* one line below the identification initials. If you have more than one enclosure, type *2 Encls.* or *3 Encls.*, as the case may be. If the enclosures are significant, list their identifications:

2 Encls. (Contract GN439, 2 copies)
(Proposal No. 54-68, 5 copies)

Carbon Copy Notations

If you wish copies of your letter sent to anyone other than the addressee, place the designation *cc:* in the lower left-hand corner of the carbon copies. If you want the addressee to know of the distribution of the copies, have the notation and the initials or full identification of the individual receiving the copy typed on the original. The carbon-copy notation appears two lines below the enclosure data:

HMW:sr
2 Encls.
cc: ELB

Postscript

The postscript is gaining favor in business correspondence. It is used to add extra emphasis to some particular item or to give additional information. A *P.S.* penned by hand gives a more personal touch to the letter. The postscript is placed two line spaces below the identification line or two line spaces below any other notations that follow it. The initials *P.S.* are often used, but are lately being omitted because they are considered unnecessary.

SECOND AND SUCCEEDING PAGES

Most letter messages can be fitted and framed attractively and neatly on a page; some longer messages may require a second or additional pages. When a letter carries over to succeeding pages, use plain, white bond paper, the same quality as the letterhead sheet. Some organizations have preprinted second-page stationery. Begin the first line of the second and succeeding pages at least one inch from the top of the page. The name of the addressee should appear at the left margin. The page number, preceded and followed by a hyphen, should be centered and the date should be at the right—forming the right-hand margin. Sometimes these elements are single-spaced, one below the other in block form on the left margin. For appearance purposes, the second and succeeding pages should contain at least three lines of text (although five lines are preferable), and should never divide a word at the end of a page. Left- and right-hand margins should be the same as those of the first page:

-2-

Cornell Aeronautical Laboratory, Inc.   August 11, 1968

or

Cornell Aeronautical Laboratory, Inc.
Buffalo, New York   14221
Page 2
August 11, 1968

THE ENVELOPE

Unless a secretary or mail clerk performs the task of opening the mail, the first thing the receiver notices about your letter is its envelope. The envelope deserves as much care as the message it contains. Assure its accuracy by checking it against the address printed on the letterhead of the company to which you are writing. The elements of neatness and attractiveness are just as important in typing the envelope as in typing the letter. There are two standard sizes of envelopes in use today:

9½ x 4-1/8 inches
6½ x 3-5/8 inches

The larger size provides greater ease in folding the letter sheets. You fold 8½ x 11 inch stationery to fit the larger envelope as follows.

1. Place the letter face up on the desk before you.
2. Fold the bottom edge of the paper up one-third of the length of the paper. Then run your hand over the fold to crease it.
3. Fold the top edge of the paper down over this fold to within one-fourth of an inch of the crease.
4. Place the letter in the envelope and insert the second fold first.

For the smaller size envelope:

1. Place the letter face up on the desk before you.
2. Take the lower edge of the letter and fold it up to within one-fourth of an inch of the top edge. Run your hand over the fold to crease it.
3. Fold the letter to the left one-third of its width. Crease the fold.
4. Fold the left third of the letter to within one-fourth of an inch of the right fold. Crease the fold.
5. Place the letter in the envelope, inserting the last fold first.

Addresses on envelopes should be typed (before the letter is inserted) in block style, double-spaced if the address is in three lines, and single-spaced if the address contains more than three lines. Since optical scanners will be used to read addresses in the near future, postal authorities recommend a standardized format and placement of addresses on envelopes. The zip code is always the last item in the address. There should be no printing after it or below it. The post office preferred format is:

Mr. James Madison
3025 Theresa Street
Arlington, Virginia   22207

The address should always be contained in the bottom 2½ inches of the envelope, with the zip code at least one-half inch from the bottom. There should be not less than two nor more than six spaces between state and zip code.

Foreign countries are typed in capitals. Use no abbreviations and no punctuation at the end of the line in the address. The person's name, title, and the name of department (if applicable) precede the street address, city, state, and zip code. Special delivery or registered mail should be stamped or typed in capitals several spaces above the address. If an Attention Line is used in the letter, it should be typed on the envelope in the lower left-hand corner. If a

Personal notation is desired on the envelope, it should appear in the same position as the Attention Line.

Mr. Nathaniel Q. Wade
Senior Vice President
Mohawk Division
Security Industries
Main and L Streets
Dayton, Ohio 45409
PERSONAL

The Buckland Company
2018 Fourth Street
Berkeley, California 94710

ATTENTION: Personnel Department

THE MEMORANDUM

The memorandum or memo is a much used form of communication in business and industry. Until recent times its use has been restricted to interoffice interdepartmental, or interorganization communications. However, it is being circulated with greater frequency out of the originating organization. Its forma is similar to that of the letter, but its tone is impersonal. Most organizations have preprinted memorandum forms for interoffice or interdepartmental communications. Its format has been highly conventionalized, although details vary from one organization to another.

Memoranda are effective communication instruments for the following situations:

1. To circulate information.
2. To make a record of policies, decisions, and agreements.
3. To transmit information of record.
4. To provide a summary of a meeting.
5. To keep members of an organization or group posted on new policies.
6. To report on an activity or situation, etc.

A memorandum should consider one subject only. The subject is stated in the Subject Line. The primary purpose of the memo is to save time for both the reader and the writer. Amenities and courtesy are sacrificed for conciseness. Although the memo form emphasizes brevity, it maintains the *you* attitude of tone and strives for the reader's good will. While the body of the memo is similar to the body of the letter, its other elements have a specialized format. The paper used is often of cheaper quality than that used for correspondence sent to persons outside the organization.

The letterhead, if one is used, is simpler in design and less expensively printed. Sometimes the heading MEMORANDUM or INTEROFFICE CORRESPONDENCE is preprinted at the top of the page, following the heading identifying the company and department originating the memo. The date usually appears in the right-hand portion of the sheet. To save paper, half-sized sheets are used for short messages. Instead of an inside address and salutation, the following three lines are used:

TO:
FROM:
SUBJECT:

In place of the complimentary close, there may be a signature of the writer; or the writer may either sign or write his initials following his name in the "From" line. The memorandum has been borrowed from the practice of military correspondence. If the memorandum is a long communication, it may be organized into a number of sections and subsections. Memoranda are frequently the most convenient mechanism to convey reporting situations.

(Sample)

TO: H. A. Breckenridge   November 4, 1968
FROM: Larry Jones
SUBJECT: Breakdown of Costs for Publishing a Revised Catalog Within Limits Prescribed at Meeting of November 2, 1968

Cost Factors		Amount
1. Engineering	2 man months	$3600.00
2. Editorial Supervision	6 man weeks	2400.00
3. Illustrations	10 man weeks	2000.00
(20-23 new photos, paste-ups, etc.)		
4. Materials		300.00
5. Composition		1000.00
6. Offset Printing		7285.00
7. Covers		1500.00
	Total	$18,085.00

cc: HCN
    WEW

52 Principles and Fundamentals

FORM CD-121 (11-63) (PRES. BY A.O. 206-10)

UNITED STATES GOVERNMENT

U.S. DEPARTMENT OF COMMERCE
NATIONAL BUREAU OF STANDARDS

Memorandum

TO : Henry M. Rosenstock
Mass Spectrometric Data Center

DATE: April 25, 1966

In reply refer to: 201.01

FROM : Herman M. Weisman
Office of Standard Reference Data

SUBJECT: NSRDS Directory Questionnaire

We are developing an inventory of the data center resources associated with the National Standard Reference Data System. The inventory is being constructed in the form of a descriptive directory. We anticipate that after the descriptive directory of the associated centers of the NSRDS is completed it will prove a helpful vehicle not only to this office but to all the centers and data projects associated with the NSRDS for defining relationships, identifying commonalties, and aiding inter and intra communication.

Attached for your information is a copy of the directory questionnaire. I would like to stop by soon to become better acquainted with your operation and to complete this questionnaire. We have filled in the entries for which we have the information in our files. I shall be calling you for a meeting at a mutually convenient time.

Enclosure

BUY U.S. SAVINGS BONDS REGULARLY ON THE PAYROLL SAVINGS PLAN

THE WHITE HOUSE
WASHINGTON

April 20, 1966

MEMORANDUM FOR THE HEADS OF

DEPARTMENTS AND AGENCIES

If Federal agencies were still operating at their 1964 level of efficiency, my 1966 and 1967 budgets would have to be $3 billion higher. These savings mean that we are getting more value for our tax dollars. It means the American people are $3 billion better off.

This makes clear why I consider cost reduction so important. It explains why I want every Government employee to think hard about opportunities for cost reduction, and why I want the best ideas publicized for all to use. A good idea from one agency should not stop there, but must be made known throughout the Government.

Some time ago I asked the Budget Director to develop a system of exchanging information about cost reduction among Federal agencies. He has prepared the pamphlet which is attached -- the first issue of a series of "Cost Reduction Notes."

"Cost Reduction Notes" describes imaginative actions which have produced savings in one agency and which carry promise of applicability throughout the Government. The ideas vary widely, but they were chosen as ones likely to be useful to agencies with differing responsibilities. By bringing the ideas together in a pamphlet which will be circulated throughout the Federal Government, we are seeking to multiply the savings already achieved.

I want "Cost Reduction Notes" to be read widely in every agency, both in Washington and in the field. I want each idea to be considered carefully. I hope that many of them can be put to use.

Lyndon B. Johnson

PART II

APPLYING CORRESPONDENCE PRINCIPLES

5
WRITING AND REPLYING TO INQUIRIES AND REQUESTS

The type of letter the technical man will write most frequently is the letter of inquiry or request. Inquiries fall into two categories: the solicited and the unsolicited. This type of correspondence makes up about half the mail received in business, industry, and government. While many inquiries deal with routine matters, organizations receiving them avoid treating them routinely because they recognize the importance of inquiries to the vitality of the organization's existence. The technical man does not buy haphazardly. He writes inquiries and asks questions about products, components, services, and systems. Routine, inadequate, confused, perfunctory, or curt answers will lose good will and a latent sale. Even when potential sales are not involved, organizations appreciate the public relations potential of the inquiry situation. Many organizations make it a practice to attempt to answer the inquiry the day it is received. Some replies require time and study; then an acknowledgement is sent, advising that the question is being studied and that a detailed answer will be mailed by a specified time. Care is taken to provide answers that will satisfy the requester.

THE INQUIRY LETTER

Inquiry letters seek information or advice on many matters (technical or otherwise) from another person or organization who is thought able to furnish it. Sometimes the inquiry may offer potential or direct profit to the person or company addressed. Frequently the recipient has nothing to gain and much time, energy, and expense to lose but, as a matter of good public relations policy, he will reply. Whether the recipient sends the inquirer the information requested often depends on the writer's ability to formulate his inquiry. A properly composed query will make the receiver want to reply and will make the job of answering easier. A poorly written request may go unanswered, may receive an answer of little value, or may start an expensive chain of letters back and forth eliciting clarification and additional information.

Inquiries are important to the economic well-being of a company or organization; many companies invest much energy and expense to invite queries. Look through any scientific and technical magazine. Every advertisement and featurette reaches out to the reader, inviting him to write for further information and details, for catalogs, bulletins, specifications, reports, manuals, and for brochures and prospectuses. For the interested inquirer, the letter to be written is simple and straightforward. The advertisement or feature has provided him with the specifics he needs. For example:

> For more complete information write to Continental Systems Control Corporation and ask for a copy of our factual illustrated prospectus, "A Community of Science," Department G.
>
> Write for the specification sheets of our new Rapid Scanning Spectrometer. Ask for Bulletin BP 206-65.
>
> Send for Catalog 77A and associated price lists for our high voltage terminations and DC kilovolt-meters.

Request and inquiry letters for these situations are short and to the point. The writer needs to identify clearly what is wanted, using the language of the invitation:

Gentlemen:

Please send me Bulletin BP 206-65 which you mentioned in your advertisement in the July issue of *Science.*

Yours truly,

Department G
Continental Systems Control Corporation
P.O. Box 209
Chicago, Illinois   60616

Gentlemen:

I would appreciate receiving more complete information you indicated was available in your illustrated propectus, "A Community of Science," about which you advertised in the January issue of *Physics Today.*

Sincerely yours,

Midwest Chemical Products
1410 Wrightwood Avenue
Chicago, Illinois   60614

Gentlemen:

Your ad in the June 8 issue of *Design News* indicated your booklet, "Products and Processes," was available on request.

I am particularly interested in epoxies—and their application in the construction of dies and molds. I would like to know the physical properties of epoxies, such as bearing load limits, heat resistance and wear resistance. Perhaps a representative of your company might call on us to discuss our possible uses of epoxies.

Yours truly,

Everett H. Lindholm

The letters are clear, concise, polite and specific; they meet the message requirement efficiently and courteously.

The unsolicited inquiry letter is one in which the writer has taken the initiative for making the request for information or advice. Often, the writer is asking for complex, difficult, and time-consuming answers. The company or organization receiving the request may not be in a position to provide a ready reply. The unsolicited query often asks much of the reader that involves his time and efforts—efforts he may not ordinarily be expected to make. You should, therefore, try to ease his burden as much as possible. Most of us do not mind being helpful, but we object to being imposed upon. To avoid creating the wrong impression, we need to plan our letters carefully to make the answering easier. The well-planned, well-formulated unsolicited letter of inquiry will have the following organizational pattern.

1. Begin the opening paragraph with a clear statement of the purpose of the letter. Define for the reader the information wanted or the problem involved—what is wanted; who wants it; why it is wanted. Do not begin with (or include in any part of your letter) an apology for making the request or for presuming on the time of the reader. (This negative approach suggests to the reader a reason for refusing the request.)

2. The second paragraph should lead into the inquiry details. Its wording should be specific and should be arranged so as to make the answering of it as easy as possible. A good technique is to state specific questions in list form. However, the request should be reasonable. The writer should not expect a busy person to take several hours to answer a long involved questionnaire. Many companies will not answer inquiries from persons who are not customers unless they know why the writer wants the information. This is especially true if the information requested is of a proprietary nature. If you need to ask for confidential material, evaluative opinions, or for any information that the reader might hesitate to reply, offer him assurance that you will handle or use the material within the limitations he may suggest.

3. The final paragraph should contain an expression of appreciation with a tactful suggestion for action. Use first person, future tense: "I shall be grateful" or "I shall appreciate." Don't use the awkward, wordy, passive, third person: "It will be appreciated if . . . " Never thank the reader in advance; this is

presumptuous and in poor taste. You might conclude your letter with a statement that you would be happy to return a similar favor or service, or that you would be glad to share any results of your investigation if the reader should so desire.

Here are some examples of unsolicited letters of inquiry.

HILL CHEMICAL RESEARCH CORPORATION
Development Laboratory
1700 Lower Road
Linden, New Jersey

May 25, 1968

U.S. Atomic Energy Commission
Division of Technical Information
Post Office Box 62
Oak Ridge, Tennessee 37831

Gentlemen:

    The Development Laboratory is actively involved in the separation of the isotopes of noble gases. We are now trying to develop bibliographies on the properties of the isotopes of these gases, with special emphasis on Argon-36, Argon-38, and Argon-40. Eventually, we will be preparing bibliographies on the isotopes of other noble gases.

    We are aware of the many excellent tables published by various activities of the Atomic Energy Commission on the properties of the gases. Do you maintain literature files or bibliographies on the gases? If so, would they be available for our use in preparing the specific bibliographies we have planned?

    Because we believe that our proposed publications will be of value not only to our Laboratory but to the scientific community at large, we are planning to make them generally available. Your assistance in providing literature references will be greatly appreciated.

Sincerely yours,

Jonathan Bruner
Research Associate

NATIONAL PHYSICAL DATA CENTER
Washington, D. C. 20234

June 24, 1968

Advanced Technology Evaluation Center
State University
2630 S.E. 4th Street
Minneapolis, Minnesota 55414

Gentlemen:

    The National Physical Data Center has been given the responsibility by the President's Office of Science and Technology for administering a system that will provide critically evaluated data in the physical sciences on a national basis, centralizing a large part of the present data compiling activities of a number of Government agencies.

    We are in the process of planning the information services operation. For this reason we are gathering information on the organization, personnel, facilities, and equipment of similar information center activities. We would therefore like to ask you for any printed or readily available information on your operations on matters such as the following items:

        General method of operation
        Type of services and charges, if any
        Handling of inquiries
        Costs
        Personnel (levels, specialities, and organization)
        Facilities (types and specifications)
        Equipment (types and specifications)
        Measures used for quality control and services effectiveness
        Statistics on number and types of documents processed
        Indexing and classification procedures
        Storage and retrieval procedures
        Most efficient techniques
        Suggestions

    We hope it will be convenient for you to send us this information. Your cooperation will contribute greatly to our providing a more efficient service to the technical community.

                                  Sincerely yours,

                                  Nathaniel E. Darr
                                  Director

NAKAMOKU & GINZBERG
Architects and Engineers
733 Marquette Avenue
Minneapolis, Minnesota 55440

March 2, 1968

National Bureau of Standards
U.S. Department of Commerce
Washington, D.C. 20234

Gentlemen:

We are interested in knowing whether the air or dust particles in hospital X-ray rooms can become radioactive by the operation of the X-ray equipment; and if so, to what degree?

The Minneapolis Building Department is under the impression that this air becomes radioactive, and it will not permit, because of this, recirculation of this air through the ventilating systems of a city hospital we are designing.

It is our considered judgement that an X-ray room is not subject to the foul odors and gases that would be encountered in other parts of the hospital, and therefore, part of the air-conditioned air can be recirculated.

We would appreciate receiving your views on this subject.

Sincerely yours,

Keiche Nakamoku
Architect

WEST POUDRE HIGH SCHOOL
Fort Collins, Colorado 80521

August 9, 1968

Union Chemical Corporation
130 East 42nd Street
New York, N.Y.

Gentlemen:

    I understand that you supply teaching materials such as a cutaway dry cell model, silicone samples, silicone-treated bricks and a small booklet entitled, "Silicones in Your Future."

    I would appreciate your furnishing me with information on the acquisition of such materials and a listing of what teaching aids you have.

    My eleventh grade science students and I are looking forward to hearing from you.

                              Sincerely yours,

                              Leslie McCurdie
                              Science Department

Smith Kline & French Laboratories
1500 Spring Garden Street
Philadelphia, Pennsylvania 19101

Gentlemen: July 17, 1968

    I would like to know how long in advance of surgery should the tranquilizing drugs 'Thorazine' or 'Stelazine' be discontinued? I have seen conflicting opinions as to the withdrawal period.

    Further, I would appreciate receiving information on preoperative and surgical procedures in patients receiving long-term therapy with 'Thorazine' or 'Stelazine.'

    I would appreciate receiving this information at your earliest convenience.

                              Yours truly,

                              E. S. Tyman, M.D.

COLORADO STATE UNIVERSITY
FORT COLLINS, COLORADO 80521

February 10, 1964

The National Business Machines, Inc.
Data Systems Division Development Laboratory
Yonkers, New York

Dear Sirs:

The report in the current issue of <u>Electronics</u> of your work on optical switching leads me to hope that you may be able to provide an instrument to solve my problem.

My research requires me to illuminate a sample ca. 1 inch inside an electron spin resonance spectrometer cavity through a hole 1/4 inch in diameter. The incident spectrum must be rich throughout the visible and up to ca. 1,000 mu. The source must be constant to within $\pm$ 2% during the time the sample is illuminated. I must provide repeated light and dark cycles going from full dark to full light and full light to full dark in less than 0.5 msec. Typically, the cycle is a 2 second one -- one second light, one second dark. However, there are occasions when the cycle may be of several minutes duration and it would be best to have a system provide indefinite light and/or dark times. I would certainly consider a system which provided either rapid change from dark to light or light to dark. However, I would much prefer to have one system give me both. The required intensity may be estimated from the fact that I presently get sufficient intensity from a 550 w. Sylvania K300 zirconium arc lamp collected by a f/1.8 lens system. This causes a reading on a Hoffman silicon cell (120 CG) placed in a thin glass cell in the sample position of ca. 10 ma.

In the present arrangement, the beam is ca. 1/4" in diameter, with the extreme rays making an angle of ca. 150° to the axis and, as mentioned above, collected by an f/1.8 lens system. My shutter takes 6 msec. to go from full dark to full light as monitored at the sample position.

I can, if necessary, junk my entire present arrangement in order to get faster transitions from light to dark.

Any suggestions which you might make would be very much appreciated.

Sincerely,

Stanley Stoller
Molecular Biology Laboratory

REPLIES TO INQUIRIES AND REQUESTS

It is the set policy of most organizations to answer all letters received promptly, courteously, and fully. Whenever possible, they grant requests ungrudgingly and graciously. Many organizations (especially in government) receive so many requests for information that, in the interest of efficiency, they use form replies in routine situations. The forms serve either as transmittal letters for the accompanying printed information or as referral to more pertinent sources of reply. A typical form reply is shown below. The inquiry is courteously acknowledged; the use of the form reply is explained; and the appropriate box is checked to indicate that the information is attached or that it may be obtained elsewhere. When a request must be refused, it is done tactfully. Most requests for information can be granted. Make sure you understand the question or problem. To avoid the possibility of overlooking any point in the inquiry, read the letter carefully and underline the points or questions asked. If any part of the inquiry is not clear or is ambiguous, make a note of it and ask for clarification in your reply. A reply to a misunderstood question may result in frayed feelings and, perhaps, in disasterous consequences.

Answer the inquiry promptly, concisely, and courteously. If a delay is necessary, acknowledge the letter with an indication of definite information as soon as it can be gathered. Direct replies should begin with a friendly statement indicating that the request has been granted, or granted to the extent possible. Then provide the complete and exact information that was requested, including whatever explanatory data that might be helpful. Refer specifically to any parts of the enclosures that are pertinent. If part of the information wanted cannot be provided, this fact should be indicated next and should be accompanied by an expression of regret and an explanation of the reasons why the complete information cannot be given. Include any additional material that might be of value. There is no better way of making a reader feel that his query has been welcome than by inviting him to return for any additional help or information he may need. End your reply with that kind of courteous offer.

Here are some examples of answers to inquiries.

66 Applying Correspondence Principles

FORM NBS-237
(Rev. 4-6-67)

ACKNOWLEDGMENT OF REQUEST OR INQUIRY

U.S. DEPARTMENT OF COMMERCE
National Bureau of Standards
Office of Technical Information and Publications
Washington, D.C., 20234

Date: _____ File No. _____

Thank you for your interest in the National Bureau of Standards (NBS). Unfortunately, the large number of requests and inquiries we receive each day prohibits our answering each of them personally.

☐ The publication you requested may be purchased for $_____ from the Superintendent of Documents, U.S. Government Printing Office, Washington, D.C., 20402. Remittances should be made by check or money order.

☐ The publication you requested is no longer available for distribution. It may be available in depository libraries for government publications. (See enclosed list.) A copy of most out-of-print Bureau documents may be purchased from the Photoduplication Service, Library of Congress, Washington, D.C., 20540. Give an accurate and complete identification (author, title, name of series, etc.), when placing your order.

☐ We cannot identify the publication you requested. Please verify that it is an NBS publication, and provide more complete information.

☐ The publication/information you requested is not available from NBS. We suggest that you write to:

☐ The NBS does not perform product tests for the general public. Nor does it publish consumer guides, or approve, recommend, or endorse any proprietary product or service.

☐ A Directory of College and Commercial Testing Laboratories (formerly published by NBS) is published by the American Society for Testing and Materials, 1916 Race St., Philadelphia, Pa., 19103. Price $1.50.

☐ Although the NBS does publish a number of codes, standards and specifications, we have no publication dealing with the specific subject of your inquiry. The NBS often cooperates with other Government agencies, as well as with scientific and industrial associations, in the development of codes and specifications. In general, the Bureau contributes methods of testing, data on the properties of materials and standards of measurement. The finished documents are usually promulgated by the sponsoring society or agency, e.g., The United States of America Standards Institute, 10 East 40th St., New York, N.Y., 10016; The American Society for Testing and Materials, 1916 Race St., Philadelphia, Pennsylvania, 19103; and the Society of the Plastics Industry, 250 Park Ave., New York, N.Y., 10017.

Writing and Replying to Inquiries and Requests   67

☐ NBS does not maintain a regular mailing list for announcement of publications. The BEST means of keeping abreast of the Bureau's programs and publications is through the monthly *Technical News Bulletin*. We are enclosing a complimentary copy, and a brochure describing the TNB and other Bureau publications.

☐ Federal Purchase Specifications are the responsibility of, and are available from, the Business Service Center, General Services Administration, Washington, D.C., 20407. The role of the NBS is that of cooperation in the preparation of certain of these specifications falling within the Bureau's area of competence.

NBS is not actively engaged in research in this area, nor do we have any current information which is applicable. We suggest that you write to:

☐ Society of Automotive Engineers
  485 Lexington Ave
  New York, N. Y., 10017

☐ National Council on Radiation Protection
  and Measurements
  4201 Connecticut Avenue, N. W.
  Washington, D. C., 20008

☐ American Watchmakers Institute
  608 East Green Street
  Champaign, Illinois, 61820

☐ The publication you have requested is in press, we will notify you when it becomes available.

☐ For information on *science projects* we suggest you write to Science Service, 1719 N Street, N. W., Washington, D. C., 20036.

☐ ADDITIONAL REMARKS (see reverse):

☐ Your letter and remittance have been sent to the Superintendent of Documents, Government Printing Office, Washington, D.C., 20402, for necessary action.

☐ This NBS publication is out-of-print; however, a reproduction copy may be purchased from the Clearinghouse for Federal Scientific and Technical Information, 5285 Port Royal Road, Springfield, Va., 22151:

Order by number _____

Price _____

☐ A wide range of government reports are sold by the Clearinghouse for Federal Scientific and Technical Information, a part of NBS Institute of Applied Technology (address above).

☐ Your request has been referred to the Clearinghouse.

☐ Please send your request with remittance directly to the Clearinghouse.

☐ We have no publication dealing with the specific subject of your inquiry. However, we are taking this opportunity to enclose general descriptive material on the Bureau.

☐ Complete descriptions of NBS publications are contained in the attached brochure, along with instructions for their purchase.

35378—U.S.Dept.of Comm—DC—1967

NATIONAL BUSINESS MACHINES, INC.
Data Systems Division Development Laboratory
Yonkers, New York

February 11, 1968

Dr. Stanley Stoller
Molecular Biology Laboratory
Colorado State University
Fort Collins, Colorado   80521

Dear Dr. Stoller:

    Thank you for writing us about your interest in our work in optical switching techniques as reported in <u>Electronics</u> magazine. I have discussed your problem with engineers in our development laboratories to see whether we might in any way be of help to you.

    Engineers here suggest that you try a mechanical deflection scheme using mirrors. These mirrors can be driven at high speeds by piezoelectric crystals or magnetostrictive rods. It is their opinion that electro-optic techniques are not suitable for your problem, since they are optically band-width limited.

    Thanks again for your interest in NBM. If I can be of further help, please let me know.

                                          Sincerely,

                                          Alfred P. Yates
                                          Laboratory Technical Information

U.S. DEPARTMENT OF COMMERCE
NATIONAL BUREAU OF STANDARDS
WASHINGTON, D.C. 20234

March 14, 1968

IN REPLY REFER TO:
143.01

* Mr. Keiche Nakamoku
Architect
Nakamoku & Ginzberg
Architects and Engineers
733 Marquette Avenue
Minneapolis, Minnesota 55440

Dear Mr. Nakamoku:

Your letter of March 2, inquiring about the possible radioactivation of air due to the use of X-ray equipment in hospitals has been forwarded to me for reply.

The X-ray installations generally used in hospitals which may run up to 3MeV do not have sufficient energy to cause nuclear reactions which are necessary to make dust radioactive. The X-ray machines used in diagnostic rooms do not generally exceed 100 kilovolts which certainly could not cause any material to become radioactive.

There is no reason, therefore, to reject reaccumulation of air from these rooms to the ventilation system because of the possibility of making dust radioactive. If you have any further questions we shall try to answer them.

Very truly yours,

P. Sanger
Health Physicist

NBS INSTITUTES FOR SCIENCE AND TECHNOLOGY
INSTITUTE FOR BASIC STANDARDS   INSTITUTE FOR MATERIALS RESEARCH
INSTITUTE FOR APPLIED TECHNOLOGY

UNION CHEMICAL CORPORATION
130 East 42nd Street
New York, New York

August 15, 1968

Mr. Leslie McCurdie
Science Department
West Poudre High School
Fort Collins, Colorado   80521

Dear Mr. McCurdie:

    Your letter of August 9 has been referred to this division by our Public Relations Department. We are pleased to learn of your interest in silicones for discussion in your science classes. We are sure your science students will enjoy studying how silicones are improving products and processes now a part of their everyday life and how they are making seemingly impossible engineering feats possible.

    Under separate cover, we are sending you the following silicone samples which will demonstrate a few of the characteristics and applications which have been found for silicones:

        L-45 Silicone Oil
        Silicone Rubber Tubing
        Silicone Varnished Glass Cloth
        Organic Varnished Glass Cloth
        Silicone-treated and Untreated Glass Bricks

    We are enclosing brief descriptions and instructions for use with these samples. We have attempted to devise the demonstration in such a way that the materials can be used many times without depleting your supply. When you exhaust your supply, we will be pleased to furnish you with additional samples.

    Several years ago we exhausted our supply of the mock cutaway of a dry cell battery, so we are sorry we cannot supply you with this item. However, we are sending you a booklet entitled, "The Union Story of Dry Batteries," which completely describes the construction and function of dry cell batteries of various types. If you are particularly interested in a cutaway of a No. 6 dry cell, please refer to page 35 which illustrates this item in detail.

    Further, two books on the subject of silicones which may be of interest to your students seeking further information are <u>An Introduction to the Chemistry of Silicones</u> by Dr. E. Rochow (John Wiley & Sons) and <u>Silicones and Their Uses</u> by Dr. R. McGregor (McGraw Hill Publishing Company).

    Technical literature on several of our products is also enclosed. We hope this material will prove useful to you and we wish your students success in their study of silicones.

                                          Sincerely yours,

                                          N. H. Ryan
                                          Silicone Division

Encls.

Writing and Replying to Inquiries and Requests

ENGINEERING EXPERIMENT STATION
OF SCIENCE AND TECHNOLOGY
Iowa State University
Ames, Iowa

Dear ___ Sir ✓ Ma'm ___ Heart ___ Me:

In reply to your recent missive,

___ I love you; let's get married.
___ Ditto; let's not get married.
___ Let's duel.
___ Send money, we'll close the deal.
___ Your contribution gratefully received; thank you.
___ Don't send money.
___ Send more money.
___ Keep sending money.
✓ The copies you want are available.
___ Not all the copies you want are available.
✓ The copies you want are ✓ enclosed
___ suppressed ___ out of print.
___ Copies are ___ ¢ each, or ___ different for $___.
✓ You're on our mailing list, you lucky fellow.
___ You're not on our mailing list, you lucky fellow.
___ Why don't you read ___ Playboy
___ Esquire ___ Time ___ Joe Miller
✓ ___ maps ___ palms.
✓ The trouble with the world is, no sense of humor.
___ You are the finest person I know.
___ You belong in ___ the White House ___ a museum ___ the booby hatch.
___ I've been out of town.
___ I'm going out of town.
___ I'm going to stay out of town.
___ Don't worry; I'll marry your daughter.
___ Don't worry; I'll not marry your daughter.
___ I might like you, but I have ___ politicians ___ dogs ___ cats ___ insurance men ___ children ___ grownups ___ women.
___ I'll meet you tonight ___ downtown ___ at the movies ___ at the pub ___ under the campanile.

___ Thank you for the gift; it's what I've always wanted.
✓ Thank you very much, I admire your ✓ brains ✓ beauty ✓ thoughtfulness.
___ I accept, though I am most unworthy.
___ Excuse my tears, but I'm so ___ happy ___ sad ___ worried ___ tired ___ uncomfortable ___ dejected.
___ Pardon me for breathing.
___ In addition,

_____

_____

Yours ✓ truly,
___ for laughing,
___ for loving,
___ for longing,
___ for living,
___ for giving,

✓ RLH   *R L. Handy*
___ Dr. RLH
___ Prof. RLH
___ R. L. Handy
___ R. Lincoln Handy
___ Sel. Service No. 13-57-29-54
___ Dick
___ Nero
___ Tyrone
___ Clark
___ Al Capone
___ Anonymous

72 Applying Correspondence Principles

Smith Kline & French Laboratories · 1500 Spring Garden St., Philadelphia, Pa. 19101

phone 215 LOcust 4-2400

July 20, 1968

Dr. E. S. Tyman
Suite 8B
8000 McLean Boulevard
Akron, Ohio 44305

Dear Dr. Tyman:

    Thank you for your inquiry regarding preoperative and surgical procedures in patients receiving long-term therapy with 'Thorazine' (chlorpromazine, SK&F) or 'Stelazine' (trifluoperazine, SK&F).

    I know of only one report, that of Scanlon, which mentions anesthesia after the chronic administration of 'Thorazine,' and I am enclosing it for your reference. I am sure that many hospitalized psychotics have undergone surgery while receiving phenothiazines, but until we receive published reports discussing the interaction of anesthesia with long-term phenothiazine therapy, it is probably best to discontinue 'Thorazine' or 'Stelazine' before elective surgery.

    It is difficult to say how long in advance of surgery the drugs should be discontinued. As you will note in the enclosed papers, Gold recommends drug withdrawal one day prior to surgery. However, phenothiazines and their metabolites have been measured in the urine as long as five to seven days after drug withdrawal. I would estimate that there is probably no pharmacological potentiation after 24-48 drug-free hours.

    In addition to the papers I have cited above, I am enclosing a reprint of a paper by Elliot, "Influence of Previous Therapy on Anesthesia." This author mentions the report by Scanlon, but he concludes that phenothiazine should be discontinuied before elective surgery whenever possible.

    We appreciate your writing us, and I have made a note of your interest. I will write you again if relevant studies come to my attention.

                            Sincerely yours,

                            Malcolm E. Davis, MD
                            Clinical Investigation Department

Encls.

    The series of letters that follows is presented not only as well-written models of letters of inquiry and reply but also as a heart-warming episode that actually occurred. The letters demonstrate how a sincere, direct approach, carefully planned and phrased in a friendly and courteous tone, can engender a reply in kind. The response in helpfulness goes further than is normally expected. The letters are a lesson in good human relations and effective communications.

215 Trotwood Road
Pittsburgh, Pennsylvania 15234

December 15, 1968

Smith Kline and French
1500 Spring Garden Street
Philadelphia, Pennsylvania 19101

Gentlemen:

    I have been a stockholder of the company through a custodial account since I was 9 years old, and have always been interested in your brochures and letters describing different product development and research; therefore, I am aware that we do a good deal of research on growth development and metabolism, and make thyroid products.

    This year I am studying biology and have become very interested in science. I now hope to make it my career.

    For my biology project this year, I would like to do a "live" experiment, rather than just the usual charts and reports alone. I have about two months in which to complete my work. If the experiment is not a long one, I would prefer to do a "trial run" before doing the one for class.

    Since my mother would probably not be too keen on allowing me to keep mice at home, I wonder whether you could suggest an experiment that I could do with tadpoles or frogs (perhaps you have a better suggestion). I thought about showing accelerated development and growth through use of thyroid or other hormones.

    Neither of my parents is scientifically inclined, and we are not permitted to discuss our particular project with our teacher in advance. The other two top students in my class are the son of a pathologist and the daughter of a dentist; and, although we get the same grades in tests and weekly lab work at school, I know I will have some real tough competition on this project.

    I will be looking forward to your reply, along with any information, reference material, etc. which you can suggest for experiments. Please be sure to include the amount and frequency for the administration of any drug or drugs suggested.

    If there is a charge for any supplies or material, will you please send them with a bill, and I will send a money order immediately, since I already have some Christmas money. I am very anxious to get started.

    I will appreciate any help you can give me, and you have my wishes for a happy holiday season.

                                        Very truly yours,

                                         Lorna Marron

P.S. If my ideas are not practical for home experiments, would you be kind enough to send other suggestions?

74 Applying Correspondence Principles

Smith Kline & French Laboratories · 1500 Spring Garden St., Philadelphia, Pa. 19101

phone 215 LOcust 4-2400

December 23, 1968

Miss Lorna Marron
215 Trotwood Road
Pittsburgh, Pennsylvania   15234

Dear Miss Marron:

    Your recent letter requesting aid in setting up a science project for your biology class has been referred to me. It is certainly understandable that you feel at a disadvantage in this assignment, since neither of your parents is a scientist. However, if your well-written letter is any indication of your ability, you should do very well despite this handicap.

    Although frogs and tadpoles are generally more difficult to work with than mice or rats, the study you have suggested, involving accelerated development and growth through the use of thyroid, is a very good one.

    I am enclosing a small paperback book entitled <u>Research Problems in Biology</u>. Chapter twelve, "Regeneration in Anurans Induced to Undergo Metamorphosis by added Thyroid Hormone," outlines an excellent experiment which you may wish to pursue. I suggest you read the list of references at the end of this chapter before you begin working, and if you still need further help, you should write to me again.

    I hope you will enjoy owning this book. It contains numerous possibilities for future studies which you may wish to carry out at some later date.

    I am glad to be able to help you in your research, and I certainly would like to know the results of your experiment.

                                                 Sincerely yours,

                                                 Beth S. Carroll

                                               (Mrs.) Beth S. Carroll
                                               Public Relations

BSC:cw
Encls.

Smith Kline & French Laboratories · 1500 Spring Garden St., Philadelphia, Pa. 19101

phone 215 LOcust 4-2400

January 6, 1968

Miss Lorna Marron
215 Trotwood Road
Pittsburgh, Pennsylvania

Dear Miss Marron:

Since I wrote to you on December 23, I have shown your letter to a number of people in this department and in our Research and Development Division. You will be pleased to know that they were as impressed with your letter as I was, and some of our research scientists offered me suggestions for you.

Although SK&F cannot send you thyroid preparation and tadpoles directly, these can be purchased at most laboratory supply houses. There are a number of reputable houses you could contact; we suggest the following:

>   Miss Bernice Smith
>   Arthur H. Thomas Company
>   Third and Vine Streets
>   Philadelphia, Pennsylvania

You should purchase desiccated thyroid preparation which can be administered to your tadpoles by simply putting it in the water in which they live.

We would like to extend an invitation to you to visit our laboratories in the near future. Because we realize you attend school during the week, we hope you can arrange to visit SK&F and write to tell us what date would be convenient to you. At that time you will see our Research and Development facilities and be free to ask any questions which may have arisen concerning your own experimental work.

If a trip to Philadelphia is not possible, please feel free to contact us if you need further help with your research. I look forward to hearing from you soon.

Sincerely yours,

Beth S. Carroll

(Mrs.) Beth S. Carroll
Public Relations

BSC:mab

January 13, 1968

Mrs. Beth S. Carroll
Smith Kline & French Laboratories
1500 Spring Garden Street
Philadelphia, Pennsylvania

Dear Mrs. Carroll:

You are an absolute "angel" for having taken such an interest in my problem and for helping me so much. I promise to prove worthy of it and to keep you posted.

Thank you so much for the book. It will be very helpful in future experiments, as well as the one which I am about to do.

I have not begun to assemble the materials for my project thus far, because I had a stubborn virus infection, and had to miss more than a week of school. Somehow it seems to take three weeks to make up the work and tests missed during one week's absence. I am also busily preparing for mid-year exams in my five majors. These get tougher every year, as the colleges become more and more selective.

I am very excited about your invitation to visit the Smith Kline & French Labs, and am certainly looking forward to such a visit, but unless the labs are open on Saturday, I'm afraid I would have to put it off until Spring vacation. How far in advance would I have to let you know?

Since I have not begun my project, no concrete questions pertaining to the experiment have come up. There are several "nagging little questions" that do come up before beginning, and that is whether to work with only three or four tadpoles and perhaps take pictures during the different stages of metamorphosis, or whether to work with a dozen or more, preserving one at each important state of development? (Of course, this would be in addition to written reports.) If it is suggested that I preserve them, what would be the best way? If one of your scientists can also suggest the best concentration of thyroid in which the tadpoles should be placed in order to greatly accelerate the metamorphosis, it would save me considerable time (and, no doubt, a considerable number of tadpoles).

May I thank you again for your interest? I realize that what you have done has taken considerable time from what must be a busy schedule, and it is much appreciated.

If the questions I have asked take too much additional time and effort on your part, as well as that of your researchers, please do not bother with them, because I do not want to become a nuisance. Please do let me know the days and hours that your laboratories are open, since I would not like to miss such a marvelous opportunity.

Gratefully yours,

Lorna

Lorna

Smith Kline & French Laboratories · 1500 Spring Garden St., Philadelphia, Pa. 19101

phone 215 LOcust 4-2400

January 20, 1968

Miss Lorna Marron
215 Trotwood Road
Pittsburgh, Pennsylvania 15234

Dear Miss Marron:

I was glad to learn that you have recovered from your illness and that you are about to begin your science project. One of the scientists in our endocrinology laboratory had offered to answer your questions and so I put him to work on your last letter. Here is the material he gave me:

Because you may have some problem with tadpole mortality, you should work with at least twelve tadpoles for any one experimental treatment. You can photograph the animals at each stage of development. However, if you prefer to preserve them, you will need some help from your teacher in using formalin.

Tadpoles can be grown in small bowls of dechlorinated tap water which can be prepared either by boiling water or by letting it stand for twenty-four hours. For a start, about fifty milligrams of thyroid in two-hundred milliliters of water should be a good dosage. You might try the experiment with several different levels of thyroid and also with elemental iodine.

Feeding thyroid to tadpoles is a simple matter. Thyroid is not soluble so you can mix it into fresh, dechlorinated water and immerse the tadpoles for twenty-four hours without food. After this period the animals may be transferred to water which contains food such as "baby food" type oatmeal, chopped raw liver or spinach. This thyroid procedure should probably be repeated at several intervals.

Two methods may be used to measure tadpoles. You can record either total decrease in body length or increase in length of one of the limbs. Measurement may be carried out by putting the animals, one at a time, in a shallow, transparent container such as a Petri dish, placed on a sheet of graph paper.

The tadpoles should be sorted out by this measuring method at the beginning of the experiment to insure that initially they are about the same size and stage of development. One word of caution, if you begin treatment with tadpoles that are too young, they will not respond to the thyroid.

Since Smith Kline & French does not operate on Saturdays, we will expect a visit from you during spring vacation. Please let us know what date would be convenient for you. We hope, by the time of your visit, your experiment will be nearing completion or, at least, well underway.

Unfortunately, I will be leaving SK&F at the end of January to have my first child. Though I hope to be able to meet you on the day of your visit next spring, all correspondence after January 29 should be directed to Miss Nancy Hays, who will be taking my place.

We are all looking forward to meeting you.

Sincerely yours,

Beth S. Carroll

(Mrs.) Beth S. Carroll
Public Relations

BSC:na

REFUSING A REQUEST

Many times we receive requests we cannot grant. Sometimes the request or inquiry is inordinate; sometimes the request is moderate and within the bounds of the situation, but still we cannot grant it. In either case, if we care little or not at all about the good will of the asker, we can dispose of the request with an easily composed, stereotyped statement:

> I regret to inform you that the policy of this organization forbids us to furnish the kind of information you seek.

It is relatively easy to be harsh and brusque—so easy to antagonize and to develop a source of ill will towards us and our organization.

It is much more difficult, but almost always necessary, to write a letter that says "No" to a request and that still retains the good will of the requester. Regardless of how extravagant the request may seem, the effective and intelligent approach is to refuse the request tactfully, with courtesy and friendliness, rather than harshly. We need to show the reader that we understand his problem, and we must be sure he understands ours.

Behind most refusals are good, logical reasons; usually, the reasons can be given. Most refusals have a job of education to do—informing the readers of some of the circumstances of which he is probably unaware that dictate the refusal. Frequently, it is possible to offer an alternative, perhaps not as attractive as the original request but still one that "compensates" for the refusal.

An effective refusal letter usually has the following elements.

1. A beginning statement that makes the inquirer feel that his request was welcome.

2. A review of the situation with explanations of the circumstances or reasons.

3. The refusal of the request (expressed or implied).

4. A suggestion of an alternative or of a possible other source where the inquirer may obtain the needed information or services.

5. A friendly close—usually an offer to be of service on other matters in the future.

These elements are present in the following examples.

AMERICAN INFRARED SPECTROGRAPHIC SOCIETY
2323 Connecticut Avenue
Washington, D. C. 20234

October 17, 1968

Mr. B. J. Cowlishaw
Technical Specialties Department
Haverstraw Paper Company
Logan, Utah 84321

Dear Mr. Cowlishaw:

Thank you for writing about your interest in the spectra described in the recent Chemical & Engineering News story.

The infrared spectra you requested are on keysort cards, and for the most part are more than ten years old, probably very much outdated. Because we have a limited number of sets available, the Society has decided to donate them to educational institutions on the basis of financial need. For example, if Utah State University should request a set on the basis that it cannot afford to obtain up-to-date sets of spectra from presently active sources, we might be able to make up a set available to the university.

If requests from educational institutions do not exhaust our limited supply, we shall make these sets available to companies such as yours on a first come first serve basis. Your letter has been placed on file for this eventuality.

Sincerely yours,

D. S. Cook
Executive Secretary

ASSOCIATION OF SCIENTIFIC INSTRUMENT MANUFACTURERS
40 Wall Street
New York, N. Y. 10005

August 18, 1968

Professor W. H. Glubish
School of Engineering
Iowa State University
Ames, Iowa

Dear Professor Glubish:

We are glad to learn from your August 2 letter that you believe our *Electronic Instrument Design Handbooks Series* would be an invaluable aid in the preparation of your students for the engineering profession. Others in the teaching field have expressed the same view, and this is one reason we are eager to extend ourselves as far as possible in making sets available to universities.

We wish there were a direct means of making a set of the Design Handbooks available to schools without cost. Because this ambitious research and publishing venture ran far beyond $250,000 in cost, we have felt it possible to give these books only to the member companies who support the Association with their dues—and even they are asked to pay a nominal amount for the binders and revision sheets.

You may be interested and pleased to learn that as of July 1 of this year we inaugurated a program under which we are suggesting to our member companies the desirability of donating sets of the Design Handbooks to educational institutions of their choice, at the special educational price of $97.50. A number of ASIM member companies have availed themselves of this opportunity, and we believe others will do so in the future. We can't very well suggest generally that institutions like yours select one or more alumni who are highly placed in the scientific instrument field and suggest such gifts to them, but this might be a course you would wish to consider.

If we can be of any further help in this matter, please let us hear from you.

Sincerely yours,

Peter A. Boucher

Peter A. Boucher
Executive Director

ORDER LETTERS, ACKNOWLEDGMENTS, AND QUOTATION LETTERS

The purchasing-department form and the purchase-order form in industry, government, and other organizations are replacing the order letter (see a typical purchase-order form below). Nevertheless, the technical man is occasionally called upon to write an order letter. As in the letter of inquiry, the order letter should include certain basic and specific items of information in order to avoid confusion, costly errors, and feelings of ill will between the writer and the seller. To ensure efficient and expeditious handling of the purchase, the order letter should include the following elements.

1. Make a specific offer to buy.
2. Describe the item completely; provide specifics on the name of the item, catalog number, model or stock number, quantity desired, and descriptive details as to material, size, color, style, and quality.
3. Indicate the unit price or the estimated total price, if appropriate.
4. Indicate payment information (C.O.D., charge, check, cash, draft, money order, or discount).
5. Provide necessary shipping instructions: the complete name and address of the buyer (and of the consignee, if any); when the item is wanted (by August 31); how the item is to be sent (parcel post, express, freight); and special instructions as to packing or specific destination.
6. Provide information on acceptable substitutes.
7. Provide credit references, if open account accommodations are desired.

You may make changes from or add to the elements listed above as circumstances dictate. Here is a typical order letter:

Sears Roebuck and Co.
Philadelphia, Pennsylvania 19132

Gentlemen:

Please send me the following surveyor's supplies via express prepaid. All are listed in your Spring 1968 Catalog. The items are:

Catalog No.	How Many	Name of Item	Price	Shp. Wt.
99P46021L2	1	Transit Level	$294.40	48 lbs.
99P46218L2	1	Surveyor's Compass	12.79	1 lb 2 oz.
9P40001	1	Plumb Bob	2.39	10 oz.
99P461134	1	Engineer's Rod	21.95	8 lbs.
			$331.33	
		Sales Tax @ 3%	9.94	
			$341.27	

Also enclosed is a check for $350.00 to cover total cost and shipment by Railway Express. I shall need these instruments by no later than April 7. Should there be any problems about my receiving the complete order by that date please contact me at once.

Sincerely yours,

Dana J. Trigle

82   Applying Correspondence Principles

1.L,S,G	2. Method	3. Code	U.S. DEPARTMENT OF COMMERCE National Bureau of Standards PURCHASE REQUEST	4. Date	5. Purchase Order No.
6. Contract No.			7. Request No.	8. Working Div. & Sec.	9. Project No.
10. Object Class			11. Work Order No.		
12. Name(s) of Suggested Vendor(s) (If more room is needed, use reverse)				13. Deliver Articles or Services To:	14. Date ☐ Desired ☐ Required
~	~	~	~	15. Room No. and Bldg.	16. Ship Via
Telephone:				17. Telephone	
18. Invitation No.			19. Date of Bid	20. F.O.B. Point	
21. Gov't B/L No.			22. Disc.	23. Delivery	Days
24. Item No.	25.	Description of Articles or Services		26. Quantity	27. Unit
				28. Unit Price	Estimated Cost
					29. Total

30. I certify that the material and/or service requested above is properly chargeable against the project indicated

(SIGNATURE OF APPROVING OFFICER)

31. Total Estimated Cost .... $

CHARGE ACCOUNT

32. Approved By:

(SIGNATURE) (DATE)

33. Material Accepted-Payment Authorized By

(SIGNATURE) (DATE)

34. Material Received By:

(SIGNATURE) (DATE)

35. Voucher No. 36. Schedule No.

37. Division Pick-up ☐ Supply Pick-up ☐
    Delivery By Co. ☐

38. For Use of Agent Cashier

Cash Advanced $

Signed

Paid Receipts $

Cash Returned $

Signed _____ AGENT CASHIER

SUBMIT IN TRIPLICATE TO PROPERTY MANAGEMENT SECTION. FILL IN BLOCKS 4 THROUGH 31. DO NOT WRITE IN SHADED BLOCKS.

(Continued on Reverse Side)

FORM NBS-10 (7/65)

84 Applying Correspondence Principles

NBS 10 (7/65)	Item No.	Description of Articles or Services (State fully)	Quantity	Unit	Estimated Cost	
					Unit Price	Total

TO BE COMPLETED BY PROCUREMENT SECTION	39. Names of Bidders	40. Addresses of Bidders								

ACKNOWLEDGING ORDERS

When orders are routine (such as a weekly order of so many nuts and bolts, made on a Monday and delivered invariably by Thursday), acknowledgments are not necessary. In most other circumstances, customers wish to know that their orders have been received, understood, processed, and are being sent. Acknowledgments, therefore, are not only desirable but are necessary. Acknowledgments confirm the order as received and as it is listed for shipment. Receipt of the letter allows the buyer to check his transaction and to inform the seller of any discrepancy or conflict between what was ordered and what is being shipped; in this way, problems and additional costs can be avoided. Problems in meeting delivery schedules or in filling the exact items requested can best be dealt with in the acknowledgment letter.

The acknowledgment letter may contain the following elements.

1. An opening paragraph that tells the buyer that his order has been received, is appreciated, and is being processed for shipment (or explains what is happening to the order).

2. Statements (sales talk) perhaps describing the worth of the purchased items to confirm that the buyer made a wise choice.

3. Offer of help with special problems or applications.

4. Statements asking for clarification if any item is not fully described or is erroneously described.

5. Statements explaining delays or necessary revisions or substitutions for any (or parts) of the ordered items. In these circumstances, approval of the changes should be requested.

6. An expression of appreciation for the order.

A typical letter of acknowledgment is shown below.

Dear Mr. Stokes:

We are pleased to confirm your order for a Lektromatic Calculator model 14B

List Price	$715.00
Federal Excise Tax	42.90
3% Sales Tax	22.71
	$780.61

Your order is being processed today from our Toledo distribution plant. It should reach you Air Freight no later than November 3rd, two days before your required date.

We are confident that as the years pass, you will look back upon your Lektromatic Calculator as one of the wisest investments you have made.

This letter is written not only to express our appreciation but to offer our services as a consultant on any calculating problems you may encounter. We would like to place at your disposal our files of Lektromatic Methods, available at no charge. Just call us.

Your new Lektromatic carries a one year guarantee, and you have at your call a highly skilled, factory trained service representative to assure the continued performance of your Lektromatic.

Please feel free to call on us at any time.

                                        Sincerely,

                                        Leo X. Orkner
                                        Sales Manager

THE QUOTATION LETTER

The quotation letter is a reply to an inquiry about prices. It capsules salient descriptive matter because complete description and specification of the product is usually given in accompanying bulletins.

Dear Mr. Stokes:

We appreciate the opportunity to provide the quotation on our Lektromatic Calculator Model 14B.

Full descriptive material is included in the attached bulletins. May we call attention to some of the features of the Lektromatic 14B which have made it, we believe, the outstanding calculator buy in America today. These exclusive features include:

—Three dial audit proof for greater first time accuracy
—Simultaneous pushbutton multiplication for greater speed
—Complete carriage carry-over to insure 100% accuracy
—Automatic division line-up and division decimal
—Automatic repeat and non-repeat
—Flexible single-key depression.

Additional advantages in the Lektromatic are *low price, low maintenance,* and *extra long life.*

The price of the Lektromatic Model 14B is:

	$715.00
Excise Tax	42.90
Sales Tax	22.71
	$780.61

Delivery: Immediate, off shelf
Terms: Net 30 days
Guarantee: One year—no charge

We look forward to receiving your purchase order.

                                        Sincerely,

6
SALES AND PROPOSAL LETTERS

Unless the technical professional is in the sales, promotion, or advertising department, he is unlikely to be directly concerned with writing a sales letter. However, scientific and technical people are often consulted to help promote or sell products and processes with which they have been intimately involved in development or with which they may be fully knowledgeable. The consultation or aid often results in a sales letter. Millions of letters selling technical and scientific products, processes, and services are written daily. Sales writing has probably influenced the technical man directly or indirectly in his own evaluation of products, materials, and services. Therefore, it is appropriate for the professional to be knowledgeable in the technique of sales correspondence—especially since (authorities on correspondence say) all business letters are, in principle, sales letters.

The sales letter has been defined as a letter whose chief purpose is (1) to effect (or help to effect) the sale of products, material, and service; (2) to effect the acceptance of an idea or approach; or (3) to promote a company, organization, group, or individual. Selling may be defined as the art of persuading or inducing a prospective customer to buy the article or thing to be sold. The sales letter is not the chief instrument in the general consumer market place; however, it is a logical medium of communication in the technical market place. This results from both the nature of the product and the nature of the customer.

Mass media are more effective in reaching customers of the consumer market because of the heterogeneous character of the customers. They can be reached quickly, easily, and effectively, and the selling message becomes saturated by advertising on television, radio, in the newspaper, and in magazines. Detergents, toothpastes, breakfast foods, tobacco products, tires, drugs, and automobiles are used and needed by everyone. The media reaching the greatest number of people are therefore used.

Technical products and services are of a specialized and complex character. The mass appeals used to sell consumer products are inappropriate because the technical customer buys the products or services not for his personal needs but for his work needs. Technical products are distinguished from products used for personal consumption by the following differences.

1. *Unit of sale.* The unit of sale of technical products is usually greater and the item is generally heavier, larger, more complex, and costlier. Faith in the product frequently depends on the faith in the manufacturer and on his technical capability to produce a technical product of merit.

2. *Customers are usually well-informed.* Technical purchases are made by men who are frequently expert in their knowledge of the product and in what they expect it to do. The technical customer may have at his disposal a well-equipped laboratory and skilled technicians to test the materials submitted. The motives that influence the buying of technical goods are rational rather than emotional.

3. *Several buyers may need to be convinced.* In many instances, several persons at succeedingly higher administrative levels may have to be satisfied before the sale of a large and expensive item is consummated.

4. *Services required.* Some instrumentation and machinery require consultation and supervision in installing and in the training of the customer's operating personnel. Also, as changes in design and improvements are developed, the addition of equipment may require attention from the manufacturer; repair services, too, may be supplied on a contract basis or may be included in the original sales contract.

5. *Role of the salesman.* Because of the factors mentioned above, the most important instrument in the selling process of the technical product is the salesman. The sales letter (as does advertising) lubricates the selling process and builds acceptance of the company that makes the product. Every salesman knows it is easier to sell a product already familiar to his prospect than it is to confront his prospect with an instrument or system (which might need explanation and interpretation) with which he is altogether unfamiliar.

Viewed in this light, the role of the sales letter is to introduce the product to a potential buyer or to introduce detailed literature describing or explaining the product so that the potential buyer will invite a salesman to call. The sales letter is an aspect of the advertising medium. It is used for a selected audience known to have an interest or a potential interest in the product or services. Its primary job is to educate the prospective consumer or user and to induce him to invite the salesman for a personal explanation of the benefits and advantages of the item being sold. One of the most common and important uses of the sales letter, then, is to convert an inquiry into a sale. Persons querying a company about any of its products or services will receive in response not only the information wanted but also a direct suggestion that a fuller and more competent reply may be obtained by the visit of a company representative. The representative, then, is in a direct position to help the inquirer meet his problem through the utilization of the company's products or services. Often a series of successive sales letters follow the lead offered by a query. The letters may continue over a period of time to keep the communication channels open for the eventual sale.

Thus, in the technical situation, sales letters may be used for the following purposes.

1. To make direct sales. (The larger, more complex, and more costly the unit, the rarer is the direct sale by letter.)

2. To encourage inquiries about products and services and to obtain leads for salesmen.

3. To announce and gauge reaction to new products and services.

4. To reach out-of-the-way prospective customers and to strengthen weak territories.

5. To help create good will.

STRUCTURE OF THE SALES LETTER

To write an effective sales letter you must understand its components and how they are interrelated. The structure of the sales letter is fundamental to any letter aimed at persuading the reader to take a favorable action. The sales letter is constructed in steps to produce the following reactions in the reader.

Step 1. Attracting the reader's attention.

Step 2. Creating interest in or desire for the product or service.

Step 3. Convincing the reader that your product or service is the best, and that now is the time to buy.

Step 4. Motivating the reader to act.

These four steps are basic. Although some effective sales letters may devote a paragraph to each, often the divisions are not clear cut. Frequently, the components shade into each other, because the development of the succeeding step flows directly from the preceding step. For example, it might often be difficult to determine where the step for arousing desire ends and the step for instilling conviction begins. The four steps will be discussed in detail so that you may better understand the structure of the letter, but bear in mind that the components in the final letter are fused into an even whole. To gain a fuller understanding of the construction of the sales letter, you might examine some of the advertisements in your technical magazines. Analysis will show the same four structural steps in operation:

1. Attention.
2. Interest (or desire).

3. Conviction
4. Action.

Attention is attracted immediately by layout, by color, by illustrations, by size of head (type), by catch phrases, statements, or questions. Interest is aroused by describing the desirable performance characteristics or parameters, the economy of the product, or by the profit potential its use provides. Conviction is instilled by the use of statistics, graphs, testimonials, case histories, test results, guarantees, and the like. Action is stimulated by suggestions for some easy effort such as "Call collect"; "Send the stamped addressed postal card for a salesman to call"; and "Fill in the coupon below for further information."

THE BEGINNING–ATTRACTING THE READER'S ATTENTION

How many sales letters do you get a week? How many of them do you read? How many do you begin to read and then throw into the wastebasket? Most of us have developed a resistance to the blandishments in sales letters. Yet all of us like to receive mail and, despite a built-up incredulous attitude to sales letters, we do approach (at least) the beginning sentences of each letter we receive with some interest. The writer of the sales letter must take advantage of this inherent initial interest to attract and compel the reader to follow his message from the opening to the end. Since the audience of the sales letter is a selected one, the appeal should be to the professional interest and working need of the reader. You must grab hold of his attention at once; if the opening fails to do this, the whole letter fails, and it is relegated with junk mail into the litter basket.

There are a number of proven ways of catching the reader's attention. A study of the ads in the technical magazines of your field will quickly identify these devices. Among them are those given below together with some examples of the technique.

1. *Stating a significant fact* directed explicitly toward the needs of the reader. Frequently the statement is set off at the top of the letter similarly to the headline of a news story, or ad, or as the Subject Line of the letter:

    a. "When you need a cushioning material to withstand temperature extremes between -100°F and 500°F, you need National silicone sheet or sponge."

    b. "You can't get Batman on our TV screen, but you can get a lot of excitement."

    c. "Napped CORFAM® used in critical polishing operations performs 500% better than other materials."

2. *A slogan, proverb, epigram, or quotation* is often effective to catch hold of your reader:

    a. "If you're renting more computer than you need, it's like burning money."

    b. "The shortest time between 2 points is a curve! We can prove it with the DK®-2A Automatic Recording Spectrophotometer."

    c. "The uncommon cold. We've been giving it to industry since 1932."

    d. "There once was a PDP-8
        that lacked a peripheral mate.
            A module or two—
            Some rice and a shoe—
    How simple it is to relate."

3. *Asking a question* frequently proves to be a most successful device to arouse the reader's curiosity and attention. Psychologically, he pauses and reads further to find the answer. (You must show, however, that you understand the reader's needs in constructing the reply. The answer should lead the reader to conclude that your product or service is the one that will meet his needs.) Here are examples:

    a. "Would you be interested in a completely functional X-ray diffraction laboratory if the cost was well under $10,000?"

    b. "Is your equipment service squeezing your profit?"

    c. "If someone gave you a handful of beads, what would you do with them? We've given away thousands of handfuls of glass beads. We can't begin to tell you the many uses our customers have for them."

    d. "How much space would you need on a 19-inch rack to cram in seven individual tape monitoring CRT displays, if each had a usable screen area of 2-3/4″ X 1-1/8″?"

4. *An anecdote* is a frequently used attention-arousing device in a sales letter. To be effective, it not only must be entertaining but also must be pertinent to the message of the letter:

    a. "From the chronicles of Araby comes a packaging tale calculated to jar the imagination. Ali Baba, you know, lived a rich life, courtesy of those storied 40 thieves. One day, they discovered that their loss was Ali's gain. To stop this till-tapping, they hid in huge jars, one chassis per package, and had themselves delivered to Ali's home as oil for the lamps of Baba. This clever low-density disguise gained them access to the palace, but egress was something else again. A slave girl found them out and did them in—very messy."

    b. "Since the J & J Soft Pretzel Company of Pennsylvania, N.J., added a new 'twist' to their operation, their production has more than doubled. The new twist is a conveyer belt made of Du Pont Armalon fluorocarbon resin coated fabric. 'We really had a problem,' states J & J president Walter R. Reach. 'Two out of every five pretzels were all right. The rest were sticking to our metal conveyer belt and the bottoms of the pretzels were being ripped up.'"

5. *The central selling point* of the product or service may be stated first and substantiated in the subsequent paragraphs:

    a. "You save 2 ways with Thompson Rolled steel rings:
        (1) You save material, because less is needed.
        (2) You save machining, because Thompson rings are contour-rolled."
    b. "Dependability, Uniformity, Availability are three reasons why *the* support for gas/liquid chromatography for over 10 years has been *Chromosorb*®!"
    c. "No wrinkles, no cracks, no blowouts! No place for leaks in the Johns-Manville heat exchanger gasket."

6. A beginning frequently used with good effect in sales letters is the *courteous command:*

    a. "Break out of the hum-drum habit if you need a drum switch. Take a new look at Cutler-Hammer."
    b. "Don't fasten your future on weak threads. Be positive with Klick Fast Blind Rivets."
    c. "Be choosy. The best reels for computer tape have aluminum hubs and winding surfaces. The best way to get them is to ask for them."

7. *The use of conditional clauses* are very effective beginnings in sales letters because they can arouse the reader's curiosity. Conditional clauses state a condition or action necessary for the validity or occurrence of the statement of the message. The word *if* is the most common conjunction used in conditional clauses. Others are *whether, in the event that, on condition that, provided that,* and *in case.*

    a. "If sterilization is a factor in your product or process, you need Amsco Sterilizing Systems."
    b. "If you get no thrill out of running a still... try Nanograde Solvents for GLC."
    c. "If the heating system you want hasn't been built yet, build it!"

CREATING INTEREST IN OR DESIRE FOR THE PRODUCT OR SERVICE

Having secured your reader's attention in the opening of the sales letter, you must now establish an interest in or a desire for your product or service.

You do this by focusing on the specific qualities of the service or on the particular specifications of the product that meet your reader's needs or desires. This second step of the sales letter—usually taking the greatest amount of space in the letter—tells the reader how your product or service meets his requirements. You do this by explaining and describing the product, how it works, or how it is made. Not all specifics are given, but you give the fundamental ones that make your product or service unique or desirable. Particulars should be left to accompanying literature, specification sheets, or to a visit by your sales representative. In the consumer-product situation, appeals to the reader are subjective—they are on an emotional or psychological plane; in the technical situation, the appeal to the reader is objective, geared to logic, to his knowledge, and to his requirements.

CONVINCING THE READER

The function of the third step is to show that the claims made for the product or service are valid or true. This part of your letter must convince your reader that your product or service meets his needs. Claims and statements in the second step must be substantiated by fact, logic, or expert testimony. In the technical situation, an effective means for producing conviction on the part of your reader is to set up a standard and then demonstrate that your product or service meets or (better yet) excels this standard. Other effective devices for convincing the reader are: (1) trial offers that give the reader an opportunity to use the product or service; (2) samples that may be provided free or at a nominal cost; (3) test results indicated in the letter or accompanied in attached literature; (4) guarantees; (5) statements of savings and economies; and (6) enclosures of additional literature in the form of specification sheets, case histories, test results, tables, reports, drawings and diagrams describing structure, performance characteristics, and the like. References to enclosures must be skillfully and specifically made in the letter. Frequently the sales letter's sole purpose is to motivate the reader to examine the particulars of the enclosure. The material attached provides the decisive details on the item or services being sold.

MOTIVATING THE READER TO ACT

Steps 1 to 3 in the sales letter will be successful only if the reader takes the action you want him to take. In the consumer situation, an actual sale may be the desired action. In the technical situation, the desired action is usually unlikely, because the item or service is too costly and complex to be sold by letter alone. The chief purpose of the sales letter is to open the potential customer's door for the flesh-and-blood salesman who will sell the product or service.

Realistically, then, the action desired in the technical sales letter is not immediate purchase but to induce the prospect to write for additional information, to ask for a catalog, to ask a salesman to call, or to come to the salesroom for a demonstration. Obtaining the desired action may involve these steps: (1) offering an inducement; (2) asking for definite action; and (3) making the action easy.

Below are several sales letters demonstrating the four major steps in the construction of an effective sales letter. Each step is identified for purposes of study.

Step 1. Attracting attention of the reader

Dear Mr. Chilcott:

Did you know that the CHN Analyzer proves its performance 100 times *every day*?

Every day more than 100 microchemists use the F & M Model 185 Carbon Hydrogen Nitrogen Analyzer to perform elemental analyses whose accuracy is well within the accepted allowable error of ±0.3%. Some have reported a distinct improvement in the reliability of their CHN analyses since they have turned to the 185. Others have already paid the 185 the ultimate compliment by sending us a repeat order.

Step 2. Creating an interest in or a desire for

An important reason why the 185 is beginning to make a deep impression among microchemists is its performance under difficult circumstances in the past year.

At the recent International Symposium on Microchemical Techniques, for instance, F & M gave a live demonstration of the 185 and promptly calculated results as follows: 0.01% error for carbon, 0.3% error for hydrogen, 0.04% error for Nitrogen.

At the Pittsburgh Conference, in a room filled with as many as 15 visitors, F & M performed on-the-spot analyses of 18 samples submitted by the visitors ... with results within the microchemist's 0.3% allowable error for all three elements.

In a series of 52 samples of all types, submitted "blind" by interested microchemists across the country, the overall standard deviation (σ) of duplicate analyses performed on the 185 was 0.15% for Carbon, 0.12% for Hydrogen, and 0.19% for Nitrogen.

Step 3. Convincing the reader

Some of the credit for the remarkable performance capability of the second-generation F & M Model 185 CHN Analyzer is due to recent design improvements: a two-stage furnace for optimum sample combustion *and* reduction temperatures, an improved single-column gas chromotographic system and an automatically timed combustion cycle ... all of which result in even more reliable analytical data than with its predecessor ... at the same speed advantage of 4 to 8 times over classical methods.

96 Applying Correspondence Principles

Step 4. Motivating action

Not the least amazing part of the 185 story is its price: $4,995.00 including balance and recorder (f.o.b. Avondale, Pa., U.S.A.). For full information, call the Chemical Instrumentation Sales Representative at one of the 41 Hewlett Packard Sales offices in the U.S. Or write me at Hewlett Packard, F & M Scientific Division; Route 41, Avondale, Pennsylvania 19311. I shall be happy to answer your questions or provide any further details.

Sincerely yours,

Step 1. Attracting the reader's attention

Dear Mr. Pritchard:

If you see our offset duplicator and buy it, you get to wear a medal proclaiming: "I lost $25."
Otherwise, we will.

Step 2. Creating an interest in and a desire for

It's a medal for heroes, sort of.
We'll pin it on ourselves if you buy any other duplicator after you see the Fairchild-Davidson.
Because we'll have handed you $25 toward the purchase of the other fellow's machine. A noble, if carefree, gesture.
But if you buy our machine, you win the medal. To commemorate the way you lost $25, but won the war for flawless duplicating.

Step 3. Convincing the reader

What are your chances of being decorated?
Well, look at it this way. More than 50% of the companies who see our demonstration and who buy an offset duplicator, buy ours.
Our only real problem is that, since we're relatively new to office duplicating, not enough businessmen know us. Hence our $25 gambit.

Step 4. Motivating action

If you aspire, look for the Fairchild-Davidson listing under "Duplicating Machines" in the Yellow Pages. Or get in touch directly with me at 516-AN 6-5200.
If I don't answer on the first ring, please be patient. I may be busy counting my medals.

Sincerely yours,

Step 1. Attracting attention

Dear Mr. Cummings:

The Russians finally admit they don't have all the answers.

Step 2. Creating an interest in or desire for

This really happened.
During a recent tour of Pittsburgh's Deeter-Richey-Sippel architectural firm, a group of Russian technicians spotted a 130 Electronic Calculator by Friden.

Sales and Proposal Letters 97

At first they couldn't believe it *was* a calculator. It looked too good.—It worked too fast. And it didn't make any noise.

Step 3. Convincing the reader

The head of the delegation asked to try it. He entered his first factor and saw it immediately appear at the bottom of the cathode ray display tube. As each new factor was entered, the previous figure moved up to the next register (this eliminates worksheets and keeps intermediate calculations available for instant use).

There was no noise because the 130 has no moving parts. Solid-state components just don't go "clickety-clack."

In the end, the Russian got his answer in milliseconds—much faster than he could index his problem.

They were genuinely impressed. One comment (which we will cherish forever) was that they had nothing like the 130 in Russia.

Step 4. Motivating action

Maybe Russia did invent baseball and chewing gum. But it took Friden to invent the perfect answer to the world's figurework problems.

Before you buy a calculator, see why the Russians were impressed. For a demonstration of the 130, just call your Friden office or write Friden, Inc., San Leandro, California. We have Sales and Service throughout the world.

Sincerely yours,

Step 1. Attracting the reader's attention

Dear Mr. Anspach:

How would you like to rent a mass spectrometer for three full months for $794.—And then get $715 of that back? May we tell you of a procedure for putting a mass spectrometer in your laboratory for a full three-month period for a maximum cost to you of $794. Honest. (And the *minimum* cost can be even lower than that.) So if you have ever wished to try mass spectrometry, this may be the time to begin.

Step 2. Creating an interest in or a desire for

First, the instrument in question. It's the MS-10—a small, accurate mass spectrometer for qualitative and quantitative analysis of gases. And despite its low basic cost (only $5290 if you'd rather buy than rent), the instrument's level of performance and accuracy doesn't have to apologize to anyone. This is a true *analytical* mass spectrometer. It is the most easily operated mass spectrometer with procedures that are quickly understood, learned, and remembered. You don't really need a specialist in mass spectrometry to use this instrument. Also, the MS-10 withstands any amount of inexpert handling; it is virtually impossible to damage. (Else why should we rent them?)

Step 3. Convincing the reader

The MS-10 was initially introduced in 1960 by AEI (the world leader in mass spectrometry). Since then, hundreds have gone into service throughout the world and, quite literally, there are now more *MS-10's* being used than any other mass spectrometer. By far, as a result, the MS-10 has been applied to more end uses than any other piece of equipment. The implication of this: the chances are good that, whatever your application, someone has already applied the MS-10 to it.

And comprehensive technical information on the use of the MS-10 for these many applications is available to you. Because of this, you usually don't have to start your work from scratch with an MS-10.

There are many accessories and vacuum systems available for the MS-10. Consequently, you need to choose only those elements which fit the MS-10 to your specific requirements. These accessories—and every subsequent MS-10 development—are designed for easy adaptation to all existing instruments. The MS-10 will never become obsolete.

Step 4. Motivating action

Now, we'll happily rent you an MS-10 for three months for a total cost of $794. Then, should you decide to buy it within the three month period, 90% of your rental fee ($715) is credited to the low purchase price of $5290.

We might as well admit that people buy the MS-10 after they've tried it. So we love to rent them. We've obviously learned that it makes sense to put the MS-10 on the road to sell itself. (To short-circuit this sequence, why not skip the rental and consider buying the MS-10 immediately?)

For more on the instrument, write for bulletin M1AC8. For more on renting the instrument, write for bulletin RM1AC8. For more on both, write for both.

Sincerely yours,

PROPOSAL LETTERS

Any communication that attempts to sell an idea, concept, piece of equipment, a complex system, or anything else is a proposal. The memo you write to your boss justifying the purchase of a new pencil sharpener or suggesting a more effective system for scheduling production of the gizmo your company is producing is as much a proposal as is the 2400-page document for developing the vehicle that might land a man on the moon and bring him safely back. The proposal delineates a problem and lays out the essential groundwork or directions for its solution. The "solution" is the statement of the proposed work, *how* it will be done, *where,* and *when.* The purpose of a proposal is to communicate clearly facts about a proposed program, plan, or technical design and, at the same time, convince the reader that this plan or design is clearly superior to those submitted by competitors. The proposal is a sales tool. The successful proposal is one that convinces the prospective customer that he should invest his money in your ideas, products, or services.

The proposal, in its usual structure, is a formal document. Yearly, in our economy, it is a multibillion-dollar source spring of new and continued activities. It represents the combined efforts of management, sales, scientific (or technical), legal, accounting, and publications personnel. The large formal document resulting from these combined efforts is beyond the scope of our topic: the principles and practices of technical correspondence. However, we are concerned with the general principles of proposal procedures that utilize the correspondence format. Specifically, we are concerned with the query letter [often known as the Request for Bid (RFB), the Request for Quotation (RFQ), or the Request for Proposal (RFP)], the transmittal letter forwarding the proposal, and the proposal in letter form.

THE REQUEST FOR BID

When an organization requires work from outside its own structure or requires some noncatalog, specialized equipment, material, or system, it locates a qualified source for the required work or equipment. The instrument for the seeking out a qualified source is the RFB (the request for bid). The formal document prepared in response to the RFB is a proposal.

The RFB is a notice and invitation to qualified and interested bidders to submit a proposal with cost estimates for the work and equipment required. It describes concisely the work to be done and specifies the requirements of the task or equipment and the manner in which the proposal or bid is to be submitted. If alternatives to the approaches indicated or to the specifications are to be allowed or encouraged in the proposal, this is so stated. Contractual conditions (if any) to govern the work, time schedules, and payments are clearly indicated in the RFB. Guarantees and liabilities (if appropriate) are also spelled out. The RFB becomes a governing document for an ensuing contract or purchase order. Prescribed formats and the number of copies of the proposal are also stated in the invitation to bid. Below is an example of a Request for Bid letter:

Scientific Research Institute
150 Charlemont Street
Newton, Massachusetts 02161

ATTENTION: Dr. Arnold Coblentz, Director of Research

Gentlemen:

    The American Association of Professional Chemical Engineers is interested in developing a set of computer routines for estimating the most useful properties of matter. There are numerous instances in which physical properties of materials have not been measured but can be estimated on the basis of theoretical or empirical relationships, either from the same properties of other materials or from other properties of the same material. This society is inviting bids (at no cost to the society) for the development of the computer routine which will provide estimates for the properties of compounds and mixtures in greatest use by American science and industry.

    The work will consist of the following major divisions:
- A. Physical Chemistry:
    1. Selection of estimation formulas
    2. Error estimates
- B. System Design:
    1. Executive computer program: (a) interpretation of input; (b) selection of route of compilation; (c) evaluation of estimation formulas.
    2. Supplementary program: addition of new estimation formulas or deletion of old ones.

The routines are to be written in a machine-independent language, such as one of the Fortran levels, so that they can be compiled and used on a variety of computers. They will be designed to accept requests for specified properties of materials; the computer will select the most appropriate estimation procedure from a collection of such procedures stored in its memory, and will compute the required data. For a given operation concerning an unknown property of a chemical compound, the developed computer routine should select the best method of estimation, and determine an estimated value of the desired property with an estimate of the error involved in the process.

    It is desired that the several subdivisions of this program be worked on in parallel so that the total period of effort of the project be no more than 24 months, which will include a pilot run of the developed system.

    Your proposal in three copies should reach this Society no later than February 1, 1969. The American Association of Professional Chemical Engineers reserves the right to reject any and all proposals; this invitation does not constitute an obligation to proceed with the project, although this Society intends to do so.

    We request that you formally advise the Society of your intention to bid as soon as possible but no later than December 2, 1968.

                                               Yours very truly,

                                             Hiram P. Tittle
                                           Executive Director

THE PROPOSAL

Most proposals—the reply to an RFB—are formidable documents and are beyond the scope of this book. It is necessary, however, to outline the essential elements of the document so that you may have a better understanding of the remaining two items to be discussed in this chapter.

Formal proposals contain the elements indicated below. The sequence of the components is not fixed; other topics may be added as the situation requires. When logic dictates, elements are combined.

The *Introduction* answers the *why* the proposal is being written. Within this background information is presented a statement of the problem, the purpose and significance of the situation, the historical background of the problem, and the scope of its boundaries. Definitions governing approach and work to be performed are indicated. The introductory section is important because it indicates to the supporting agency whether the proposer understands the problem to be investigated.

The *Technical Presentation* is the "guts" of the proposal. This section defines and describes the proposal plan for doing the work:

1. You should explain clearly what you propose to do. You need to indicate also what specifications you will meet, the scientific or technical work you will do, and the equipment, system, product, or report you will deliver.

2. You need to respond to the Invitation to Bid, point by point, item by item.

3. If you have exceptions to any of the requirements, you should explain why they will be necessary, or why it will be advantageous to the supporting agency to make the exceptions.

The *Technical Description* is the creative contribution by the proposer for solving the problem. Your method of attack based on theory, state-of-the-art, and fundamental principles is explained, justified, and substantiated. Diagrams and other visuals to supplement the text matter should be used. New approaches and modifications of standard techniques should be explained to convince the reader that the problem will be solved. If proprietary items or processes are used, these items need to be clearly indicated. Factors of this kind may help convince the reader you may be in a unique position to solve the problem.

The proposer's *capabilities* in personnel, facilities, and experience are presented. Biographical data on personnel who will do the work are included.

Programming of the work of the proposal should be carefully blocked out or charted. To be included are scheduling of work, personnel, facilities, and time.

Cost Schedules should include wages and salaries, equipment (capital and expendable), and miscellaneous costs such as, for example, travel, communications, and outside services. Pertinent overhead costs, administrative costs, and profits or fees should be clearly indicated, and justification for them should be presented.

PROPOSAL TRANSMITTAL LETTERS

The formal proposal is transmitted from proposer to requester by a transmittal letter. It is a formal device of record, which not only presents the proposal but also provides last-minute emphasis of the items in the document that the proposer considers especially significant. The letter refers to the Request for Bid and calls attention to the factors deemed of interest or of importance. Sometimes the letter will summarize very briefly the technical presentation and the cost matters. Finally, by implication or direct statement, the letter expresses appreciation for the opportunity to bid. This is usually a natural and graceful way to end the letter. The letter below is a sample for study.

Scientific Research Institute
600 Merrick Road
Lynbrook, N. Y. 11563

June 10, 1968

U. S. Atomic Energy Commission
Washington 25, D. C.

Attention: Director, Technical Information Division

Gentlemen:

In response to your request for bid of September 10, 1967, we are pleased to submit our attached proposal, Technical Memorandum AEC No. 2, for the development of a classification system for Nuclear Science.

Scope of Work

The scope of work to be performed is covered in detail in the attached Technical Memorandum.

Performance Period

The work described will require a performance period of three (3) months.

Reports

A preliminary report will be submitted at the end of the performance period. The final report will be submitted four weeks thereafter.

Consideration

It is estimated that the total price for this work, including professional services and disbursed expenses, would not exceed $22,000. We estimate that the cost for our professional services, based on our established rates, will be $18,410, which represents over five man-months of effort, during a performance period of three (3) months. Disbursed expenses are estimated at $3,590.

Government Proposal Information

Additional information concerning Scientific Research Institute and conditions pertaining to this proposal are contained in the attached, "Scientific Research Institute Government Proposal Information," which is hereby made part of our proposal.

We in Scientific Research Institute look forward to the possibility of undertaking this interesting work in your behalf. Any additional information about this proposal of our Company will be furnished immediately upon receipt of your request.

                        Very truly yours,
                        SCIENTIFIC RESEARCH INSTITUTE

                        Laurence F. Harper
                        Contract Administrator

Copies: Twelve (12)
        Attachments
            Government Proposal Information (12)
            Technical Memorandum AEC No. 2 (12)

THE LETTER PROPOSAL

Proposals on relatively simple matters, problems, equipment, or services require simple documentation, and are frequently and efficiently housed in a letter format. The components of a proposal (previously identified) can readily and effectively be constructed within a letter as the following two examples show.

                        Colorado State University
                        College of Engineering
                        Department of Agricultural Engineering
                        Fort Collins, Colorado 80521
                        December 8, 1968

Mr. R. S. Pebley, Manager
Home Construction Products Division
Rocky Mountain Iron and Steel Crop.
820 Yuma Street
Denver, Colorado 80204

Dear Mr. Pebley:

I have your letter inviting a proposal on testing nails for withdrawal resistance. We would be glad to conduct the required standard test on one or more types or sizes of nails. We would conduct the test on 20 specimens of each type or size of nail in accord with paragraph 8 of the "Gypsum Dry Wall Contractors International, Recommended Performance Standards for Nails for Application of Gypsum Wallboard," dated June 30, 1960.

We would begin the tests immediately on the signing of a contractual agreement between your firm and the Colorado State University Research Foundation. The costs given below include the preparation of a report and an overhead cost. Additional specimens can be tested for considerably less than the first one.

Our proposal for testing the first size, 1¼ inch No. 14 gage nail is:

1. Salaries:
   Professional Engineer, 0.6 month $600.00
   Clerical, 8 hrs. at $2.00   16.00

2. Equipment and Supplies:
   Strain gage elements   25.00
   Lumber and supplies   25.00

3. Report duplication:
   50 copies, 3 pages each   20.00

4. Annuities, 6% of Salaries   37.00

5. Overhead, 36% of Salaries   221.76

Total (First Test) $944.76

Tests on additional types or sizes will be made for 40% of the cost of the first test or $377.90.

We would be glad to have the opportunity to be of service to your firm and will look forward to hearing from you.

        Yours truly,

        Norman A. Evans
        Head of Department

        Tek-Doc Corporation
        700 North Henry Street
        Alexandria, Va. 22314

        February 20, 1968

Dr. H. A. Renslow
Director of Engineering
Southern Electronics Company
1200 Lee Street
Greensboro, N. C. 27920

Dear Dr. Renslow:

We are pleased to submit our estimate for preparation of Handbooks and Parts Catalog to cover the Radar Test Set and Spectrum Analyzer which your company is designing for the Bureau of Weapons.

The accompanying material represents our carefully considered estimate of the numbered pages, line drawings, schematics and photographs which will be required to adequately and properly describe and illustrate the various Handbooks and Parts Catalog in conformance with specifications as described in the attached cost analysis forms.

Handbooks and Parts Catalog will be prepared in accordance with respective specifications and in accordance with format specification 5474A. Repro pages for all Handbooks will be typeset. The Parts Catalog will be prepared on IBM electromatic typewriter for reasons of economy. The cost shown for the Illustrated Parts Breakdown is based on full MIL-STD-125 descriptions. In the event that your contract specifically authorizes MIL-STD-125 with Attachment No. 1, the cost shown will be modified downward.

Costs, as shown on the attached cost analysis forms are:

Handbook of Operating Instructions	$3770.50
Handbook of Service Instructions	7895.48
Handbook of Overhaul Instructions	4112.37
Illustrated Parts Breakdown	8923.77
	$24,702.12

We believe it would be desirable to make a second analysis of this job, particularly on the Parts Catalog, at the time the design of the equipment has been frozen in order to determine, within reasonable limits, the actual number of parts requiring description.

We guarantee compliance with all specifications and their requirements; we will perform liaison with the reviewing agency and can assure you that our staff of writers and illustrators will only require a minimum of time from your Engineering Department insofar as obtaining assistance is concerned. We will relieve your organization of many responsibilities connected with preparation and processing for approval of the Handbooks and Parts Catalog.

Our company is a government approved facility with military security clearance. Your consideration of our services in performing this job is appreciated.

Sincerely,

Stanley Jordan
Vice President & General Manager

Enclosures:

Cost Analysis Forms:
 Operation Handbook
 Service Handbook
 Overhaul Handbook
 Illustrated Parts Breakdown

7
SPECIAL-PURPOSE LETTERS

This chapter examines a miscellaneous grouping of correspondence that the technical professional may find necessary to write in doing his work. Some of the situations covered call for specific approaches; others are so devoid of complexity as to call for no special comment other than the reminder that the general principles of correspondence are applicable.

COMPLAINT AND CLAIMS LETTERS

In a world less than perfect, all of us sooner or later find ourselves confronted with the results of a transaction unsatisfactorily consummated. Through human frailty or misunderstanding, critical services are disappointingly rendered or much-needed merchandise is consigned too late in improper quantities, shipped to Miami, Florida when the destination is Portland, Oregon, and, when the material finally arrives, it is found to be defective.

Our initial and natural reaction is anger. Fortunately, we usually have a second thought and realize that we also can make mistakes. In just such a correspondence situation the application of the principles of the *you* psychology will pay dividends. Years ago business and industry generally regarded complaints as nuisances and treated them as such. Today a more progressive attitude prevails. Organizations look at complaints as opportunities to learn what is wrong with the aspects of their operation that cause customer dissatisfaction so they can remedy the situation. Progressive organizations know that a dissatisfied customer will take his business to a competitor. Properly adjusted complaints mean better customer relations and better profits.

Bear this in mind when you are the one who has been given cause to complain; furthermore, it is more than likely that the person to whom you send your complaint was not the cause of the dissatisfaction. Take up your pen with a courteous, calm, restrained but vigorous hand. Your purpose is not to give someone hell—no matter how much someone may deserve it—but to straighten out the problem or to obtain the satisfactory service or product that is due and needed. State your case factually, objectively. Accusations, invective, or sarcasm put obstacles in the way and make adjustments difficult.

Claim letters, although courteous in tone, are forthright and direct. They have the following four elements in their organizational structure.

1. A clear explanation of the situation or of what has gone wrong. Full details are provided for identification of the defective product or service. The statement should include dates of arrival or nonarrival, order number, amounts, model numbers, sizes, and colors involved. If there has been damage or breakage, this should be specified. Any other information that would help the reader check the matter should be given.

2. A statement of the loss or inconvenience resulting from the mistake or defect.

3. An effort to motivate the reader to take the desired action by appealing to his sense of fair play, his honesty, or his pride.

4. A statement indicating what the writer considers a fair adjustment. If the situation is sufficiently unclear regarding what an equitable adjustment should be, the writer should appeal for a prompt investigation and action.

Here are some examples of claim and complaint letters for study:

(a) Jordan, Smith & Marsh Laboratories
20 Bridewell Place
Clifton, N. J. 07014

Gentlemen: Reference: Our Purchase Order No. 52-94

Your shipment of our order sent with your Invoice No. 7855 consisted of 5 containers holding 278 smaller pieces. In one of the smaller pieces 3 X 500 Relaxazine Tablets, 100 mg. were received broken, in concealed damage, total value amounting to $118.14.

Since greater attention is given to claims filed by the consignor who patronizes the transportation company, we request that you file claim for $118.14 and send us credit for that amount.

Attached for your information and use is the Inspection Report and Freight Bill No. 409574 with the notation of damage reported. Should you desire any additional information, please don't hesitate to let us know. Your prompt attention to this matter will be appreciated.

Yours very truly,

SOMERSET DRUG COMPANY

Fanny Hemple
Fanny Hemple
Claim Department

108 Applying Correspondence Principles

(b) President Lyndon Baines Johnson
President of the United States of America
The White House
Washington, D. C.

Dear Mr. President:

I was greatly disturbed to read in last night's newspaper an article describing the failure of legislation concerning the proposed Study of the Metric System of measurements to pass the House Rules Committee.

I feel that such legislation is very important and necessary, particularly in our time when such giant strides are being made to build a Great Society. I feel that our outmoded and wasteful system of measurements does impede the progress of scientific development in our country. To continue to maintain the old system of measurements is to hold back the progress of the United States and to subject us to ridicule in the eyes of the world.

Since you have a great deal of influence over the Congress, I hope that you will urge the Rules Committee to bring the proposed legislation before Congress, and that you will also seek to have this bill passed into law.

I think this would be one of the changes in our country that would mark you in history as one of the great presidents. May I, in the meantime, commend you on your outstanding leadership of our nation in these critical times.

Respectfully yours,

Bernard P. Glauber
Science Teacher
Northeast High School

Sometimes a touch of humor is an effective way to ease the tension in a minor claim situation. We can look pretty silly if we treat a $2 item as if it were worth a fortune. Humor cleverly used can change an antagonistic reader into an ally. But humor can be a dangerous weapon; you may get a laugh, but not results. Be sure your humor is appropriate. If you are funny at the expense of the reader, he will resent it; sarcasm or humor in poor taste offends. The first letter example below uses humor effectively; the second letter example is clever at times, but is mostly strained and overdrawn; it does not succeed.

(a) Dear Sirs:

Your letters inviting subscribers to extend their subscriptions are excellent—a little too good, as it turns out, for me. I have just inadvertently resubscribed twice for three-year subscriptions. I hope you can cancel the second order and return the second $12.00. It is not that I don't intend to go on reading your magazine, but six years is too long a period of time to commit myself to anything (except marriage).

The first resubscription was by check dated February 12. The second was by check dated and mailed March 27. I did not become aware of this until my bank statement arrived yesterday with the two cancelled checks. That six-week interval between your two mail solicitations was too long for me to remember I had already sent in my money, but not long enough—apparently—for your records to get caught up.

Please check this matter for me. A copy of the most recent mailing cover for my subscription is enclosed.

Sincerely yours,

(b) Mr. Idiot Computer[1]
The Society of the Sigma Xi
51 Prospect Street
New Haven, Connecticut 06511

Dear Mr. Computer:

I have received your recent communication and I wish to advise you that

(1) I am a regular member of the University of X Chapter, not an associate member.

(2) On December 31, 1965, I mailed a check for $7 to one of your more-or-less human attendants to cover my dues for 1966-67 and to pay for my promotion from associate member of the Y College Chapter to regular member of the University of X Chapter. I realize that you are nothing but a conglomeration of bent wires and transistors (mostly burned out) but could you take the trouble to cash that darn check? Because you have kept it so long, my wife is getting angry at you according to the equation,

$$A = A_0 e^{t!}$$

where A_0 = anger at time $t = 0$.

I urge you to cash that check within the next $\pi/2 \times 10^7$ sec or face the prospect (probability > 1) that my wife will rip your damn transistors out not later than December 31, 1966.

(3) In view of the fact that I have paid my dues for the entire year 1966, I would appreciate receiving a copy of The American Scientist, Vol. 54, No. 1, Spring 1966. Please dispatch one via dog team, pony express, camel back, or the U.S. mails. Somehow (no doubt it was a perturbation which will not reoccur), I received my Vol. 54, No. 2 a day or two ago.

(4) I have paid my dues to the University of X Chapter for this year. Why are you billing me again?

(5) Don't you know how to add? Since the total money requested was $0.00 please find that amount enclosed.

Sincerely yours,

---

[1] Adapted from *American Scientist,* September 1966, pp. 279A-280A.

ADJUSTMENT LETTERS

Most organizations and companies are sincere and energetic in their desire to eliminate causes for complaint and have a set policy to grant adjustments when their investigation shows the claim is fair. Few companies can afford the policy that the customer is always right and no reputable firm can afford the *caveat emptor*—let the buyer beware!—policy. Claims are usually decided on their individual merits. In today's business environment, every complaint or claim, no matter how trivial, is answered promptly, courteously, and tactfully. When the adjustment is granted, the letter usually will include the following elements:

1. The reader is thanked for calling attention to the difficulty or problem.
2. He is told immediately that steps already have been taken or will be taken to correct the problem and the loss or damage will be made good.
3. The facts about the problem may be reviewed and explained. Alibis and buck-passing are considered poor form. When expedient, the writer tells the reader what actually happened.
4. In some situations, there are certain steps the reader needs to take to expedite the adjustment; these are explained and requested.
5. The adjustment is granted ungrudgingly, and the writer voices his desire to maintain good relations with the customer and expresses appreciation for his business.

In the situations where the writer must refuse the adjustment, he begins similarly on a positive tone. The reader is thanked for calling attention to his difficulty. The problem is reviewed; the facts surrounding the claim are thoroughly and courteously reexamined from the viewpoint of the decision. The adjustment is refused or only partially accepted with an explanation. The writer must show the reader that he understands the reader's problem. He must also make certain that the reader understands the situation from the writer's view.

For study purposes, here are some examples of adjustment letters replying to some of the claims and complaint letters just previously cited.

(a) Miss Fanny Hemple
Claim Department
Somerset Drug Company
400 North Michigan Avenue
Chicago, Ill. 60611

Dear Miss Hemple:

Thank you for bringing to our attention the damage of three each, 500 Relaxazine Tablets 100 mg. which occurred in our January 12 shipment.

Please use the enclosed shipping label to return the damaged merchandise to us. Upon receipt, we will make an immediate adjustment to your account.

We appreciate your sending us the inspection report as it will enable us to file claim with the carrier. We regret the inconvenience this has caused you.

Sincerely yours,

(b) Mr. Bernard P. Glauber
Science Teacher
Northeast High
Minneapolis, Minnesota

Dear Mr. Glauber:

Your letter to President Johnson on legislation for a proposed study of the metric system has been referred to me for reply.

President Johnson recognizes that the problem confronting us is how to adjust to the increasing use of the metric system throughout the world. Its solution could have, as you know, far-reaching consequences on practically every phase of national activity. We must sooner or later face and solve this problem which becomes more serious from year to year. That is why the Administration strongly and emphatically supports legislation calling for a meaningful study along the lines spelled out in H.R. 38, H.R. 1154 and H.R. 2626.

There are really two sets of problems here: one has to do with software and the other with hardware. Software problems are those whose solutions involve paperwork and training of people, such as retabulation of data and learning to think and work in terms of a different measurement system. Hardware problems are those whose solutions require changes in existing physical entities, such as machines, instruments, devices, stock sizes and standard modules. The two classes of problems are fundamentally different, and a different strategy for solution must be found for each.

Any extensive change from our present usage of weights and measures will encounter the vast momentum of existing investment and commitment based on our present customary system. There are many opinions on the diverse aspects of the problem, but nearly everyone would agree that there is not enough information on either the immediate or long-term aspects of the whole problem to evaluate or arrive at well-founded courses of action for the country.

That is why the Administration favors that a program of investigation, research, and survey to determine the impact of increasing worldwide use of the metric system on the United States be conducted under direction of the Department of Commerce. We need to evaluate the desirability and practicability of increasing the use of metric weights and measures in the United States in order to do the wise and proper thing.

The Administration certainly agrees that a study of the metric system's impact and what should be done about it is long overdue. We, too, would like to see this study get under way as soon as possible; and we are, therefore, urging prompt and favorable action on the study bills.

The President asked me to tell you he was pleased to receive your thoughts on this matter and is especially gratified by your kind words and support.

Sincerely yours,

(c) Dear Dr. Bridewell:[2]

As you well know, a computer can only do what it is programmed to do.—Would that humans might respond as well!

I shall attempt to explain some of the questions which you have raised in your June 6 letter and will take them up in the same order which you numbered them.

1. You are about to be recorded as a regular member of the University of X Chapter and not as an associate member as listed on your bill. The card which accompanied your bill stated, "If there has been a recent change of status in your membership from Associate to Full Member, it is possible that the processing of the data was not completed before this billing." I am enclosing another such card with this explanation underlined. Your Initiate Card was received in this office on April 27, 1966—too late in consideration of approximately 10,000 to process before the May 18 bill. Also, a technicality (3A below) has prevented this processing.

2. Your letter indicated that on December 31, 1965, you "mailed a check for $7 to one of your more-or-less human attendants to cover my dues for 1966-67, etc." Is it possible that this check was sent to the Treasurer of the Y College Chapter? May I suggest you contact him to have it clear the bank. We at National Headquarters have not been involved in the actual handling of this particular check.

3. We are sorry you have not received Vol. 54, No. 1 of <u>American Scientist</u>. I have mailed another to you. Routinely, we at National Headquarters ship copies prior to each initiation for distribution to the new initiates. The Y initiation was held in March, just a week before publication. We therefore sent December 1965 copies for distribution. I hope you did get yours.

In all cases of new initiates, we must await receipt of the initiate data cards before getting a Member or Associate Member into the system to receive routinely copies of <u>American Scientist</u>. Your situation is somewhat different from routine. Please let me review the facts—that is, facts as we at National Headquarters view them.

(A) The National Convention at Berkeley in <u>December 1965</u>, voted that an Associate Member could only be promoted if he were active for the current year.

(B) On <u>January 25, 1966</u>, University of X sent us the certificate order for its March initiation. Your name was on it listed as a <u>promotion</u>.

---

[2]Adapted from *American Scientist*, September 1966, pp. 279A-280A.

(C) Checking records at National Headquarters showed that you were an Associate from Y College but were <u>not currently active.</u>

(D) In anticipation of receiving current dues from University of X for you we added your record to the computer system.

(E) On March 7, X sent $3 dues for 633 listed individuals to cover 1966 (the calendar year). Your name does not appear to have been included.

(F) On April 27, Y College sent initiation and promotion fees—but only $1 was included for you—the promotion fee, but no National dues to make you active.

4. National Headquarters is not billing you <u>again.</u> I find no record of a recent billing by us of you.

I hope the foregoing may be helpful to you in excusing, or at least understanding the "Idiot Computer." Since you are still in limbo relative to the promotion, I am sending copies of this correspondence to the Secretary and to the Treasurer of the Y College Chapter. There may be some communication which has been over-looked.

Also, since we have been reporting to our National President that our new computer system has been working quite well, I have taken the liberty of sending him copies of our letters so that he may see that not everyone is delighted with our operations.

Sincerely yours,

THE LETTER OF INSTRUCTIONS

The technical professional, because of his specialized knowledge, is often called upon to teach others who have a need for his knowledge. The instruction may be direct and oral or it may be done by letter or memorandum when distance or convenience dictates. The letter or memorandum of instructions is also the instrument used by the employer to employee, superior to subordinate, or client to expert to insure that specific procedures for accomplishment of a task are carried out. If the task is a complex one, requiring diverse contributions by a number of persons using varied devices, tools, and equipment, the instructions may assume the character and format of a report or manual. Our concern here is with simpler situations that are conveniently related in letter or memorandum form.

Instructions must be framed from the viewpoint of user needs. If your reader is an expert, then your approach is technically sophisticated, geared to his expertise. If your reader is a layman, you may need to begin with basic explanatory information to bring his level of understanding up to the point where he can follow the necessary step-by-step directions.

Use the imperative mood for giving specific instructions. The imperative is the form of verb used to state commands or urgent requests. The subject of the verb is not expressed but understood to be the person commanded or requested. For sake of illustration, the explicatory material that follows is written in the imperative.

1. Put yourself in your reader's shoes; do not talk down to him nor use language above his range of experience.
2. Be careful you do not sound brusque or arrogant. (This is a danger in using the imperative. Being courteous and friendly is not inimicable to letters of instruction.)
3. Be sure your instructions are in parallel grammatical construction. (This admonition is made not for the sake of grammar but for clarity.)
4. Make your instructions complete, specific, and clear.
5. Write concisely and precisely; avoid roundabout expressions, vague or unnecessary words, and generalities.
6. Use devices such as numbering, underlining, and indentation of headings because they help to clarify the instructional procedures.
7. Be sure there is logical order in the sequence of steps in your directions; arrange your instructions in successive or chronological order.
8. Specify when your reader is to use his own judgment and when he is to follow explicit directions.
9. Be sure to include any necessary cautions or precautions.
10. Identify and list necessary tools, materials, and equipment.
11. If appropriate, specify the time or date when the task must be completed.

For study purposes, here are some examples of letters of instruction.

Mr. Michael W. Dempsey
Assistant Chief
Weights and Measures
State of Wyoming
Cheyenne, Wyoming

Dear Mr. Dempsey:

I am glad to review for you recommended Federal procedures for examination and testing of weighing devices.

You should expect your inspectors to have a full knowledge of the fundamentals of the design and operation of a weighing device. I am attaching for your information a copy of National Bureau of Standards Handbook 94, <u>The Examination of Weighing Equipment,</u> which provides an excellent background of such fundamentals. To keep current with design modifications and new models, your inspectors should study the manufacturers' latest catalogs and other descriptive literature, as well as give thoughtful study to the devices themselves.

Inspection and testing are closely allied, and at times the line of demarcation is thin. In general, inspection may be defined as that portion of the examination of a piece of apparatus conducted independently of the physical standards of weight; and testing is that portion of the examination involving the use of such standards.

Before testing a scale, it is proper and advisable for your inspector to assure himself that the working parts of the scale are in condition to function as intended. This preliminary inspection may be complete, embracing all of the elements of the scale mechanism, or it may be partial, covering only the more important or readily accessible elements. Usually, the partial inspection will be sufficient unless or until trouble develops, indicating the need for a thorough inspection.

I am listing below the more important items of inspections for common types of scales. Although the list of items under Preliminary Inspection appears formidable, the preliminary inspection ordinarily requires only a few minutes on the part of an experienced inspector. Under Inspection Following Unsatisfactory Results, the thoroughness should be dictated by the circumstances, and ordinarily a complete examination of all parts of a scale are not necessary.

Preliminary Inspection
    For general freedom from binding conditions.
        1. Examine for clearances:
           a. Around platform of built-in scales (3/8 inch to 3/4 inch).
           b. Around stock rack of livestock scales. (Rack must be mounted on the platform. Check for possible binds between gates and approaches.)
           c. Around platform and between platform and frame of self-contained scales.
        2. See that:
           a. Platforms are free to move a limited amount, and will return to normal position after displacement.
           b. Foreign material has not accumulated beneath counter scales.
           c. Stabilizing links are free.
           d. Open side of the hook of the counterpoise stem faces away from the trig loop.
           e. Weighbeam pivots are centered in loops, weighbeam is balanced, and beam action indicates general sensitiveness.
    For general cleanliness.
        1. See that there is an absence of:
        Dirt in weighbeam notches.
        Dirt in weighboom loops.
        2. Make sure there is an absence of:
           a. Rust, oil, gummy deposits, etc., on weighbeam pivots.
           b. Dirt or other foreign material on load receiving element—platform, platter, scoop, pan, etc.—and on counterpoise weights.
    For general operating conditions.
        1. Examine for:
           a. Rocking of platform, especially on warehouse and portable types.
           b. Tightness of bolts securing weighbeam pillar and shield and other exposed structural parts.

116 Applying Correspondence Principles

    c. Centering of weighbeam—front to back—in trig loop.
    d. Battered zero stop on weighbeam.
    e. Battered weighbeam poise or deformed reading edge or other index of weighbeam poise.
    f. Worn notches on weighbeam.
    g. Defaced graduation marks or figures on weighbeam or reading face.
    h. Security of balancing of material.
    i. Agreement between weighbeam or reading face indications on dealer's and customer's sides of scales.
    j. Suitability of opening in chart housing to insure readability of indications at all times.
    k. Suitability of any attachments, extended platforms, special load receptacles.
    l. Suitability of counterpoise weights in use.
2. See that:
    a. Poises on notched weighbeams are equipped with panels that fit the weighbeam notches.
    b. Springs on spring-controlled weighbeam poise pawls are strong enough to seat the pawl properly in the weighbeam notches.
    c. Dash pots on automatic-indicating scales are in proper adjustment.
    d. The operations of application and removal unit weights (drop weights) on automatic-indicating scales are positive, and that the value of the unit weights in place at any time is clearly indicated on the reading face.
3. Give consideration to:
    Probability of evidence of fraudulent manipulation.

Inspection Following Unsatisfactory Test Results

Examine any of the following elements that might produce the unsatisfactory results observed:

1. <u>Pivots:</u> For tightness and alignment, and for sharpness and cleanliness of knife-edges.
2. <u>Loops and other bearings:</u> For smoothness of bearing surfaces, cleanliness, and alinement with opposing knife edges.
3. <u>Nose-irons:</u> For evidence of movement from factory sealing positions.
4. <u>Antifriction points:</u> For sharpness and cleanliness.
5. <u>Antifriction plates, caps, and other surfaces:</u> For smoothness and cleanliness.
6. <u>Levers:</u> For alinement and level.
7. <u>Connections:</u> For vertical alinement.
8. <u>Moving parts:</u> For evidence of friction with adjacent parts.
9. <u>Cooperating parts, such as rack-and-pinion assemblies:</u> For cleanliness, smoothness, and evidence of excessive wear or deformation.
10. <u>Supporting members, such as lever stands, eye bolts, timbers, foundations, etc.:</u> For security of positioning and evidence of deformation.
11. <u>Linkages, connections, etc.:</u> For cleanliness, freedom of movement, and absence of deformation or other damage.

12. Dash pots: For frictional effects.
13. Weighbeam poises: For lost locking screws or other missing parts and for presence of foreign material within the poise.
14. Adjustable elements: For insecurity of positioning.
15. Steelyard or beam rods: For freedom of hook engagements, and for end-for-end reversal.
16. Steel tapes or ribbons: For kinks, bends, roughness, adhering foreign matter, etc.
17. Surfaces over which steel tapes operate: For roughness, deformation adhering foreign matter, etc.

I think you will find the attached NBS Handbook 94 a very valuable reference guide for your office activities. It contains a coordinated series of step-by-step examining procedures for weighing equipment and such supplementary information as field standards, report forms, weighing principles, reference material on scale construction and performance, and tables of weights and measures.

Should you need any further information, feel free to call on me again.

Sincerely yours,

Mr. Dennis Keogh:
Production Superintendent
Charles River Boat Company
11762 Sorrento Valley Road
San Diego, California 92112

Dear Mr. Keogh:

You are right, Mr. Keogh, premature corrosion at or near weld areas can be a troublesome and expensive problem. I believe, however, I can suggest an effective and inexpensive way for eliminating the problem.

When protective coatings break down in weld areas, it can usually be traced to the fact that harmful deposits formed during welding have not been fully removed before the coating. Harmful deposits commonly found near weld seams are:

1. Alkaline slag from the weld flux, which reduces the adhesion and durability of the coating film;
2. Condensed flux fumes, which produce similar undesirable alkaline conditions;
3. Oxides produced by the heat of welding;
4. Weld metal spatter.

Beads of weld spatter may be as large as 1/4 in. (6 mm.) in diameter, and their peaks are normally too high to allow adequate coverage by an average film thickness of coating. Spatter, therefore, presents vulnerable points for early rust formation.

A simple three-step pretreatment will eliminate problems caused by all four types of deposits:

1. Treat the weld with 10% phosphoric or 10% hydrochloric acid to neutralize alkalinity. Scrub the acid into the weld area with a

stiff brush. Commercial, ready-made preparations (e.g., Rust-Oleum Surfa-Etch) are also available for neutralizing the deposits, and they are easier to store and use. Be sure your workers wear protective rubber gloves, aprons, and goggles.
2. After the acid pretreatment, rinse the entire area thoroughly with fresh warm water: While the surface is still wet, remove any rust spots or oxides near the weld by rubbing with fine steel wool. Then dry very carefully.
3. You can then remove the weld spatter by sand-blasting or grinding with power tools.

Surface preparation is, of course, only the first stage of weld protection; the primer coatings do the continuing rust prevention job. For optimum performance, use only those primer coatings specifically formulated to provide maximum anti-corrosive protection for steel surfaces, including welded areas. Examples include:

a. A lead-free red metal primer (X-60), which dries to the touch in 4-6 hours and may be exposed to weather up to nine months before application of the finish coat.
b. A fast-drying (touch-dry in 30 minutes) formulation (678).

Use an intermediate coat of 960 zinc chromate primer; it will help assure long term freedom from rust. Because of its light color, zinc chromate serves as an excellent undercoat when the finish coat is also light.

Durability of the coating system will depend on film thickness. Each coat, when dry, should be at least 25 (0.025 mm.; 0.001 in.) thick but not more than 50 (0.050 mm.; 0.002 in.). You must follow the instructions of the coating manufacturer for mixing, thinning, and application of the specific coating.

In case you do not have convenient referral to sources of primer and coating preparations I am attaching some catalog sheets on preparations of our manufacture. If you have a need for any further information on steel weld corrosion problems, do please call on me again.

Sincerely yours,

THE LETTER OF AUTHORIZATION

The letter of authorization is similar to an order letter; it orders services rather than products or materials. In the opening paragraph(s) it identifies the purpose of the work authorized or commissioned, the nature of the problem to be investigated, the scope of the activity to be carried out, the direction where the solution might lie, and an indication (if known) of how the results will be used. The commissioner will specify time requirements and funds available or limitations imposed by time and money. Specific personnel required or desirable in the performance may also be spelled out. The letter may be accompanied, as appropriate, by a legal contract for the requested service. For study, below, is a historic letter of authorization.

The White House
Washington, D. C.

November 17, 1944

Dear Dr. Bush:

The Office of Scientific Research and Development of which you are the Director, represents a unique experiment of teamwork and cooperation in coordinating scientific research and in applying existing scientific knowledge to the technical problems paramount in war. Its work has been conducted in the utmost secrecy and carried on without public recognition of any kind: but its tangible results can be found in the communiques coming in from all over the world. Some day the full story of its achievements can be told.

There is, however, no reason why the lessons to be found in this experiment cannot be profitably employed in times of peace. The information, the techniques, and the research experience developed by the Office of Scientific Research and Development and by the thousands of scientists in the universities and in private industry, should be used in the days of peace ahead for the improvement of the national health, the creation of new enterprises bringing new jobs, and the betterment of the national standard of living.

It is with this objective in mind that I would like to have your recommendations on the following four major points:

First: What can be done, consistent with military security, and with the prior approval of the military authorities, to make known to the world as soon as possible the contributions which have been made during our war effort to scientific knowledge?

The diffusion of such knowledge should help to stimulate new enterprises, provide jobs for our returning servicemen and other workers, and make possible great studies for the improvement of the national well being.

Second: With particular reference to the war of science against disease, what can be done now to organize a program for continuing, in the future, the work which has been done in medicine and related sciences?

The fact that the annual deaths in this country from one or two diseases alone are far in excess of the total number of lives lost by us in battle during this war should make us conscious of the duty we owe future generations.

Third: What can the Government do now and in the future to aid research activities by public and private organizations? The proper roles of public and private research, and their interrelation, should be carefully considered.

Fourth: Can an effective program be proposed for discovering and developing scientific talent in American youth so that the continuing future of scientific research in this country may be assured on a level comparable to what has been done during the war?

New frontiers of the mind are before us, and if they are pioneered with the same vision, boldness, and drive with which we have waged this war we can create a fuller and more fruitful employment and a fuller and more fruitful life.

I hope that, after such consultation as you may deem advisable with your associates and others, you can let me have your considered judgment on these matters as soon as convenient—reporting on each when you are ready, rather than waiting for completion of your studies in all.

Very sincerely yours,

Franklin D. Roosevelt

Dr. Vannevar Bush
Office of Scientific Research and Development
Washington, D. C.

LETTER OF TRANSMITTAL

The letter of transmittal is the instrument by which a report or other formal document is transmitted to the receiver and serves as a matter of record. If the document is transmitted to someone within one's own organization, the format used is a memo. The letter introduces the report or document to the intended receiver, stating the *how* and the *when* the report or document was requested and indicating its subject matter and method of its transmittal—as an enclosure or under separate cover. Its length varies. Frequently, it is sufficient to say, "Here is the report on X which you asked me to investigate in your letter of August 11, 19___." Most transmittal letters are longer.

The first paragraph of the letter of transmittal identifies the document or report being submitted and refers to the authorization under which it was prepared. The letter may refer to specific parts of the report and may call attention to important points. Some letters of transmittal contain certain elements that appear in the document itself such as statements of purpose and scope, or they may mention specific problems encountered in making the investigation as, for example, a delay resulting from a strike or a shortage of certain materials. The transmittal letter may end with an expression of appreciation by the writer for being allowed to take part in the work of the report, or it may call attention to, and underscore, important conclusions and recommendations, as does the transmittal letter of the President's Commission on Heart Disease, Cancer and Stroke shown below.

The President's Commission on Heart Disease, Cancer and Stroke

Dear Mr. President:

I have the honor to submit the report of the President's Commission on Heart Disease, Cancer and Stroke.

The Commission was appointed by you in March 1964, to develop a realistic battle plan leading to the ultimate conquest of three diseases—heart disease, cancer and stroke—which now account for more than 70 percent of the deaths in this country. In your initial charge to us, you requested us to recommend practical steps to reduce the heavy losses exacted by these diseases through the development of new scientific knowledge and through the delivery to all of our people in every part of this great land of the precious, lifesaving medical knowledge we now possess, but fail to bring to so many stricken American families.

Grateful beyond measure of expression for this Presidential mandate, we plunged into our assigned task—confident that the toll of these three diseases could in fact be sharply reduced now and in the immediate future. During the intervening months, as we sought and received testimony from scores of leaders in medicine and public affairs, our conviction mounted that we could chart a truly national effort—calling upon the full resources of Federal, State and local governments, the dedicated members of the health professions, and our great voluntary health organizations—leading to the increased control, and eventual elimination, of heart disease, cancer and stroke as leading causes of disability and death.

This report embodies our recommendations for such a united effort by a free and vigorous people. Our stated goals are neither impractical nor visionary— they *can* be achieved if we so will it. They *must* be achieved if we are to check the heavy losses these three diseases inflict upon our economy—close to $30 billion each year in lost productivity and lost taxes due to premature disability and death.

In the early decades of this Republic, our people tended to view disease as an irrevocable and irreversible visitation from an implacable Fate. Our remarkable progress against many diseases over the past half century—the life span of the average American has been lengthened by 23 years since 1900—is vivid proof of the reversibility of any disease process.

The great engineer Charles F. Kettering once observed that no disease is incurable; it only seems so because of the ignorance of man.

We submit this report, Mr. President, in the deep conviction that its immediate implementation will not only narrow appreciably the spectrum of our ignorance, but will contribute to the saving of thousands upon thousands of American lives now needlessly sacrificed to these three deadly enemies of mankind.

Respectfully yours,

Michael E. DeBakey

MICHAEL E. DEBAKEY, M.D.,
Chairman.

8
EMPLOYMENT LETTERS

In the conventions of present-day business, industry, government, and education, the job-application letter plays an important part in introducing qualified professionals to prospective employers. Some letters achieve the introduction gracefully, effectively; others—in greater number—provide only a pedestrian introduction; and many, unfortunately, stumble and fumble gracelessly, stupidly. The problem many persons have in applying for a job is that they have little realization of and give small heed to the outlook of the reader—the prospective employer. Many job applicants, no matter how capable or experienced, are self-centered—so preoccupied with their own problems that their approach to applying for a job is unfortunate. Their letters sadly lack the psychology of the *you* attitude. In advancing their qualifications and capabilities they do not relate them to the requirements of the employer. Consequently, employers find these applications inept and inappropriate.

THE VIEW FROM THE PERSPECTIVE OF THE EMPLOYER

For purposes of insight, let us examine the job-opening situation from the prospective employer's viewpoint. Imagine that you are an employer with a vacancy for a responsible, well-paying position. You want the best qualified person you can find to fill it, so you advertise not only in your city newspaper but also in the financial pages of the *Sunday New York Times.*

By the end of the following week, the post office will have delivered more than 100 applications. You are both delighted and overwhelmed to receive so many responses. How do you find the right person from so many applicants? A scanning of the envelopes enables you to begin the weeding out: you find an envelope with the remains of the applicant's breakfast on it; further scanning reveals three envelopes with the name of your company misspelled. Anyone who is so sloppy, careless, or ignorant is obviously unsuited for the position you have open. You drop the four envelopes, unopened, into the wastebasket. The real winnowing process requires a reading of the letters. You roll up your sleeves and begin. The first letter is trim and businesslike; it begins: "Replying to your ad in Sunday's *Times,* I wish to be considered...." As you read the letters, each seems like the previous one. Nearly all begin with a participial phrase like "Regarding your ad," or "Having seen your ad," or an equivalently dull opening such as "This is relative to your ad."

After reading a dozen of these letters, you open one with a different beginning. The first few sentences are original, interesting, and pertinent. They lead immediately into the essential qualifications you want in the applicant. Furthermore, the writer is able to describe himself, his experience, and his background in dimensions that stand out. The letter clearly introduces a man with character, ability, background, and experience. This is a man you will want to interview; you carefully set aside his letter.

You return to the pile of letters. In most instances, you read no further than the first few sentences, because the letters sound so much alike. You are looking for persons such as the individual whose letter caught your interest and made the writer come alive. After the reading chore is over, you have four stacks of letters: those to be thrown away (applicants who are completely unsuitable or who are crackpots); those to be acknowledged politely (applicants with limited or very ordinary qualifications); those to be acknowledged with more care and then filed for possible referral (applicants who have good qualifications but not the specific ones you are looking for); and those—not more than six or seven—from applicants whom you want to interview to select the final candidate for your vacancy.

From this hypothetical (but typical) situation, it is evident that the reception of an application letter depends on these features:

1. *Its initial impression.* If the letter envelope is tidy and correctly typed and the letter is neat and well-framed on the page, the reader has a favorable predisposition to the letter's content.

2. *Its first few sentences.* If they arouse immediate interest and lead the reader to the applicant's qualifications, the prospective employer will continue to read. The applicant should emphasize at once his particular qualifications which he believes are most pertinent to the job.

3. *Its individuality.* An employer becomes bleary when reading scores of applications. One applicant, even though qualified, begins to sound like another. Consequently, the writer must set himself apart from other applicants if he is to be remembered. This, of course, is not easy, but we shall soon observe how it can be done.

JOB ANALYSIS AND SELF-APPRAISAL

The application letter is essentially a sales letter; the product you sell is your employability. The planning of an effective sales campaign for yourself (as with any product) requires a careful analysis of both the market and the product. You need to find the answers to these questions: What do the customers (employers) want? What do I have? Many persons, even the very experienced and capable ones, approach an application for a position from the point of view of their own eagerness for it. They forget that a prospective employer has his own viewpoint and needs. Therefore, unfortunately, the letters these applicants write do not show how their experience, training, and personal qualifications can be beneficial to the employer.

The first step in the application letter (one that begins before the letter is actually written) is self-appraisal—the construction of an inventory of your personal characteristics and physical attributes. This inventory includes the following items.

1. Education and specialized training.
2. Experience.
3. Activities; membership in professional, civic, social, or student organizations, and offices held (if any).
4. Skills, abilities, and hobbies.
5. Honors and achievements.
6. Personal data: for instance, age, health, and marital status.
7. Aspirations.

The second step is the "market" analysis—examining the position that is available. The analysis must be based on the information describing the job. This information, of course, appears in the advertisement or announcement revealing the availability of the job. A helpful technique is to itemize on a sheet of paper the position's requirements listed in the announcement. Then, in a column alongside, list the inventory of your own qualifications—your background, training, experience, and personal characteristics that are appropriate for the opening. Compare what is required with what you have. This inventory helps you understand, through comparative analysis, the qualifications and their importance to the employer. It also allows you to examine your own background more objectively in light of what the prospective employer is looking for. If you do not have what the employer wants, there is no point in applying. Save yourself time, energy, and emotional investment. Sometimes your qualifications may not be exactly what the employer is looking for, but you may have some of the desired qualifications, plus additional attributes that will be attractive to the prospective employer. There is no precise formula to gauge just how many qualifications you need to make it appropriate for you to apply for a particular job. This depends on your analysis and knowledge of the requirements of the position being advertised.

If the ad reads,

> **SENIOR ANALYTICAL CHEMIST**
> Chemist with advanced degree to work in a methods development activity. This job offers opportunity for individual expression in the development of analytical methods in areas of smoke, tobacco, chewing gum, and toiletries. Prefer man with five to ten years experience in analytical chemistry, with knowledge of gas chromatography and spectroscopy. Excellent salary and opportunities. Apply Box 24C, Richmond, Virginia.

there is little point in applying if you are a recent graduate and lack the essential ingredient of industrial-research experience in analytical chemistry and in chromatography and spectroscopy.

It is true, however, that employers look for personal qualities that no amount of schooling and work experience can provide. Industry has made many surveys to identify the reasons why employees lose their jobs. Incompetency is low on the list. Very high on the list are personal factors: absenteeism, belligerency, hypochondria, alcoholism, inability to work with others or to take orders, and the like. Personnel managers scrutinize applications for clues on prospective employees' instability, health, and emotional problems, personality disorders, or difficulties. Employers are always on the lookout for prospective employees who will be responsible, who have stability, initiative, and characteristics of potential growth. Employers not only want to hire individuals who have the personal knack of getting along well with others, but also persons who may have leadership and managerial abilities. Evidence of these factors is eagerly searched for in application letters.

In your inventory, you should be careful to note the qualities that are attractive to employers. Even though all of us have a sincere belief in our abilities, that we have social grace, and a capability of working productively with others, we must be aware that mere assertion of these traits is not enough. What is required is evidence in the form of proven achievements. If you have been elected to an office in your professional society or in a civic group, you have evidence of leadership. Recall any experience in extramural or extracurricular matters that offered challenges. Identify instances of productive accomplishments.

THE LETTER OF APPLICATION

Coverage of all seven items in the inventory (previously mentioned) could require a letter of two, three (or more) pages, full of tedious details, and the more important qualifications might be lost in the clutter. Modern practice has evolved a format to make the application letter both complete and readable: a letter in two parts. Part One is a one-page letter of about four paragraphs featuring the most significant qualifications for the position. Part Two consists of a resumé (sometimes called personal record or data sheet), which gives a detailed inventory of the applicant's background, education, experience, personal history, and references.

PART ONE—THE LETTER

The application letter is essentially a sales letter; its organizational construction has the same four components or steps.

Step 1. Attracting the reader's attention.
Step 2. Creating an interest in the "product" on the part of the prospective employer.
Step 3. Convincing the reader about the worth of the "product."
Step 4. Motivating the reader to act—to grant the writer an interview.

To attract attention, the application letter must have an opening, written in a tone of sincerity, which will arouse the reader's interest (Step 1). To create desire, the letter should describe the major qualifications of the applicant in such a way that the reader will be attracted to him (Step 2). The description, however, is not enough; there must be proof to confirm the reader's growing conviction that the applicant is qualified (Step 3). Finally, the stimulation of action is a request for an interview (Step 4).

As in the sales letter, these four steps are basic. Although some effective application letters may devote a paragraph to each step, often the divisions are not clear cut, and the components shade into each other, but the development of the succeeding step flows logically from the predecessor. This is especially true of Step 3, which needs to instill conviction in the reader of the applicant's qualifications. The conviction comes from sincerity of tone, the provision of particulars in the qualifications, and is reinforced by the details in the accompanying resumé which the letter persuades the reader to study.

Few people are hired on the basis of a letter of application alone. Its purpose is to create an interest in the applicant and to instill enough conviction that he has desirable qualifications to induce the reader to invite the writer to be interviewed.

Attracting Favorable Attention

A stereotyped opening sentence, as we have seen in our earlier hypothetical situation, can be unfortunate. Good beginnings reflect the *you* attitude; they are simple, direct, distinctive, but not odd or extravagant. Here are some examples of effective, *novel* beginnings:

> The May issue of *American Mathematical Monthly* had a feature story on IBM that convinced me that mathematicians employed by IBM have envious opportunities for professional growth and personal advancement. As a graduating mathematical statistician, I believe I could do no better than to start my mathematical career under one of your training programs.
>
> Please consider my qualifications for the Principal Industrial Engineer you are seeking for your Division to establish standards, methods and procedures for electrical and mechanical assembly. Briefly, here is what I have to offer:
>> A B.S. in Mechanical Engineering and an M.S. in Engineering Administration;
>> Eight years of short run production engineering experience in electrical and in the hard goods industries;
>> Demonstrated capability to plan and implement manufacturing standards and methods for achieving balance work flow;
>> Patience, emotional maturity, and a desire for new challenges.
>
> I am a young biologist eager to secure a position where I can earn my keep and, at the same time, learn how to be a better scientist. I understand from Dr. Paul Millsaps of your Pharmacology Department that you are expecting to add to your staff a physiologist who has a knowledge of antibiotic assay methods. May I have a moment of your time to acquaint you with my background and qualifications for this position?

In trying to be original, be careful that you do not become outrageous. No one will deny that the following openings are unhackneyed, but they defeat their purpose by their oddness:

> It has been said that leaders are made, not born. However, genetics can often lend a helping hand. Agreeing with this but at the same time recognizing the education I received at the South Dakota School of Mines and Technology along with my personal background and innate ability, I definitely feel that I am a leader. Because of this leadership ability I am certain that I could fill the Management Trainee Position now open with your company.
>
> Symbiosis is the biological term used to indicate a cooperation between two or more systems toward a mutual goal. Your company has an outstanding opportunity for an experienced structural engineer. I have an outstanding background in that field. I feel both of us have much to gain by getting together.

If you have learned of the job opening from a person whom the prospective employer knows or whose names or title will draw the employer's respect, then refer to that individual in the opening sentence. It is an effective way for obtaining the reader's attention. Here are some examples of *reference* beginnings:

> Professor W. C. Giuliani, Head, Department of Mechanical Engineering at State University, called me into his office this morning to see if I were interested in the quality control position described in the notice you sent him.

128 Applying Correspondence Principles

> The Placement Office of the American Chemical Society has informed me that my qualifications might be of interest to you for the NMR Spectroscopy position you have available in your inorganic compounds division.
>
> Dr. Roger C. Butterworth, Vice President of Engineering Research in your San Diego office, suggested this afternoon that I write you concerning an opening in your Cleveland office. He said, "Dick, because I know you have an unusual five-year educational combination of engineering and management, I think you have just the right background and, of course, the ability to make yourself quite useful to the Industrial Engineering Department in our Cleveland office. Would you be interested in dropping them a note at once?" I was delighted to hear of this opportunity and am acquainting you with my interest and background.

Because the prospective employer is interested in hearing about your qualifications, a *summary* of your most significant qualifications is often a very good way to open your letter. Here are some examples of this approach:

> My eight years' work at the York Company in decomposition of gases to form solids, in heat flow in hot substrate reactions, and in mechanical properties of refractory filaments and coatings, as well as my degrees in physics and electrical engineering, make me confident that I can qualify for developing processes for making refractory filaments.
>
> Since receiving my M.S. in Chemical Engineering at the Massachusetts Institute of Technology five years ago, I have done polymer research for Circle Chemical Corporation, concentrating on development of PVC resins. My work also included exploration of new polymerization processes, studies of the relationship between polymer properties and processing techniques, and the development and application of new evaluation methods.

Starting your letter with a *question* can be, psychologically, very effective. When anyone is asked a question, he pauses to listen. This method is especially appropriate when you are applying for an unadvertised position. Here are some examples:

> The financial pages of the *Kansas City Star* carried a story about your plans for expansion. Does your new program seek the services of a shirt-sleeve methods engineer, capable to methodize machine shop and assembly operations, set manufacturing rates, and write operations sheets? If so, please consider my qualifications.
>
> Does your laboratory need a Ph.D. biochemist with a background in microbial chemistry, enzyme chemistry or enzyme kinetics? If you do, perhaps my ten years experience in the application of enzyme systems to ecological problems may be of interest.
>
> Wouldn't that field engineer you advertised for in the Sunday *Journal* be more valuable to your company if he had as much as three and a half years' experience as a maintenance technician in the Signal Corps?

If you decide to use the question beginning, you must make certain that your qualifications answer the question you pose. If your qualifications do not answer the question you pose, very likely the question device will cause your application to be thrown in the wastebasket.

Adapting Your Qualifications to the Job

If you have succeeded in writing an effective beginning, you have captured your reader's interest. Now he will want to see evidence of your qualifications. An applicant has four major qualifications: experience, education, personal qualities, and personal history. Just which of these four you emphasize in your letter depends, of course, on your analysis of the requirements of the position. The principle is to write first and most about that qualification judged most important to the position available. If the ad reads, "Wanted—Experienced Optical Systems Engineer," obviously the employer will want to know first and most about your experience in optical-systems design. Although he will be interested to know that you were graduated *magna cum laude,* this may not be as important to him as five years' experience in designing optical and radar sensors and in analysis of command and control systems. Applicants (especially those with Ph.D.'s) often wonder whether education or experience should be presented first. The answer always is: discuss the qualification most vital to the job. If you have experience that is appropriate to the requirements of the opening, that is the most important qualification to stress; discuss it first and most. Your educational background then becomes the frosting on the cake.

If your strongest selling point is your education (which is often the case with young, graduating professionals), then this is what you need to start with. For the average college graduate or advanced-degree student, education is the strongest selling point. If your personal qualities are your strongest point, then show that you can develop the qualifications in performing the job or that you are capable of gaining the proper experience necessary to do the job. Also show that you are capable of growing into more responsible positions within the organization. After you have presented your strongest point, discuss other qualifications that are important to the position. Always be specific about details. Give the name of the company or organization for which you worked, your job title or the level of the position you held, and the duties connected with the position. If you accomplished anything outstanding on the job, tell about it.

General and ineffective: For three years I served as Associate Project Manager.

Specific: For three years I served as Associate Project Manager in the GIGA Reactor Program of the Lambeth Atomic Corporation, Bridgeport, Connecticut. I was responsible for a $4,500,000, three year experimental program in the support of the design of a new type of power reactor. This program included materials irradiation studies, critical assembly design, construction and operation, a major in-pile loop experiment and operational analysis of the nuclear power plant. Results obtained from this program aided in the sale of thirty research reactors and led to the development of a new type of pulsed reactor.

General and ineffective: I joined the company as a technician and stayed in their model ship for three years.

Specific: I joined the model shop of the research and development department of the Aerosystems Corporation, Houston, Texas. My work consisted of making breadboards, fabricating prototype models from rough engineering sketches. This required not only expert use of a soldering iron, but also machine shop tools and equipment. I was promoted to assistant model shop supervisor at the end of my third year.

In discussing your education, call attention to the subjects directly or closely related to the position being applied for. For example, a student applying for a job with a construction company might indicate his educational qualifications as follows.

My major area of study and interest was in the field of concrete structures which included such courses as Concrete Design, Prestressed Concrete, Reinforced Concrete Design, Advanced Structural Design, and Concrete Laboratory. Some further pertinent courses included Statics, Dynamics, Strength of Materials, Steel Design, Soil Mechanics, Fluid Mechanics, Water Supply, Machine Drawing, Physics, Chemistry, and Mathematics and Geometry. In my Special Problems Course in Concrete Laboratory, I designed and built a concrete canal and spillway system for the University Hydraulics Laboratory which is being used in their experimental program.

The body of the application letter, by its substantiation of qualifications, creates an interest in and a desire for the applicant.

Securing Action

The final paragraph of the application letter (the last step) has a twofold duty—to request an interview and to facilitate its granting. Very few (if any) jobs are secured on the basis of an application letter alone. The real purpose of the application letter is to arouse sufficient interest on the part of the prospective employer to ask you to come in for an interview. Proof of your qualifications is reinforced by the details in your resumé. Be sure you have invited the reader's close attention to it. This may be done in the final paragraph or may be woven into the discussion of your qualifications in the middle portion of the letter, whichever is appropriate.

Give as much care to the individuality and vigor of the ending of your letter as you gave to your beginning. Do not be hackneyed or weak; do not be impertinently brash—or timorous:

If you think I can quality for your position, I'd be grateful if you gave me an interview (negative and timorous).

I suppose it's only fair to tell you you'd better call me before the end of this week since I am considering several very fine job offers and I would want to compare yours with the others before I reach my decision (brash and impertinent).

Trusting you will grant me an interview at your convenience, I shall await your call (hackneyed).

More effective ways of closing the letter of application to secure an interview are shown in the examples below:

I shall be in Chicago from December 22 to January 2. Would it be convenient for me to come in to tell you more about myself and to learn more about the position you have available?

I believe my qualifications are appropriate for the position you advertise. I would be delighted to provide details beyond those outlined in my attached personal data sheet. May I have an interview at your convenience? Won't you write me at my home address or call me at 484-2620 between the hours of 5:00 and 9:00 P.M.?

Whether you have an opening at the present or not, I would appreciate the opportunity of meeting you so that you may better judge my qualifications for possible employment with your organization. I look forward to the possibility of hearing from you soon.

May I have an interview at your convenience? I have attached a self-addressed postal card for you to let me know when I might stop by to better acquaint you with my qualifications.

PART TWO—THE PERSONAL DATA SHEET OR RESUMÉ

The personal data sheet or resumé inventories the applicant's background, experience, education, personal history, references, and other pertinent information, organized in a readable and logical manner, and (through its details) substantiates the qualifications. The purpose of the application letter is to prompt the prospective employer to examine your data sheet in detail to see whether your qualifications are appropriate for the position available. Conviction of your fitness for the opening should stem from the material in the resumé. Even in government where the "Form 57" application blank replaces the resumé, capability in arranging one's background information in the best way to reveal experience, accomplishments, and education (achieved by the resumé) helps in the structuring of the civil service application form. In government, as in many cases in industry, the resumé and application letter precede the invitation to the formal filling out of the application form.

The resumé or data sheet organizes attractively and fully, for easy readability, the details of your background. Centered at the top of the sheet are the name and address of the applicant. In the professions or pursuits in which appearance is important to the accomplishment of the job, an application-size photo is attached to the personal data sheet. Jobs in teaching, sales, theatrical pursuits, public relations, and advertising require photographs. In most technical professional activities, a photo is not required.

Information included in the resumé is usually arranged under these headings:

> Personal data
> Experience
> Education
> References

The headings and the data under them should be arranged in the order of their probable interest to the employer and suitability to the position. If you are just out of college, list the details of your education first; if you have been out of college many years and your experience is extensive, place the experience section before the education section. It is customary to list experience in a reverse order because prospective employers are interested chiefly in what an applicant has done most recently. Identify specifically these items:

> Dates of employment
> Employer, address, and nature of business
> Position you held
> Name and title of supervisor

Indicate your job duties—the tasks performed—emphasizing the ones that required the highest degree of skill and judgment. If appropriate, indicate specialization and any duties beyond your regular assignment. Include special tools, instruments, or equipment used and the degree of skill involved. Tell how many persons you supervised, whether you held an administrative position, and to whom you were responsible. Outline outstanding results achieved. If possible, give concrete facts and figures. If you are a recent graduate, include your summer and part-time jobs. If you have worked your way through college (partially or fully), indicate this, since it is of interest to the prospective employer.

Under "Education," begin with the latest degree. List your education in reverse chronological order: college through high school. List the college courses (especially graduate courses) that are important to the job opening. Indicate all accomplishments, honors, and extracurricular activities. Include additional special-training courses that are appropriate to the position or to your field of interest, as well as any professional certificates or licenses.

Under "Personal Data," include personal and physical history such as age, height, weight, marital status, and dependents. Data on religion and racial origins are included in this section. (Most states have laws against employers asking about these matters on their application blanks, but employers usually are interested in obtaining this information. Whether you include these items depends on the laws of your state and, perhaps, on your own preference and judgment.) Memberships in professional organizations are listed under this heading; include offices held and other important contributions. Patents and inventions may be listed here also. Hobbies might be indicated only if they are appropriate to the position. However, significant hobbies might be listed if you are a recent college graduate without much back-up experience. This information could serve as back-up material for a fuller delineation of your character and personality. If you have a significant number of publications, you might provide full bibliographic detail under a separate heading: "Publications."

Personal references are important only if an applicant has no work experience. Provide not only the names of the persons given as references but also their official titles and their mailing addresses. Preferred references are persons who have supervised you on jobs or professors who have taught you the courses significant to the position you are applying for. Unless character references are requested, omit them. Be sure you have permission to list your references. Only three or four are necessary.

The resumé or data sheet should always bear a date to assure the prospective employer that the contained data are current.

Below are sample letters of application and their data sheets.

> 7262 Los Amigos Drive
> Rancho Cordova, California 95670
> July 17, 1968

Dr. Rufus H. Zelenka
Director of Research
Pan American Nuclear Research Corporation
4900 Fairmont Avenue
Bethesda, Maryland 20014

Dear Dr. Zelenka:

Yesterday I received a call from Dr. Sam Wortman, with whom I coauthored several papers on thermal decomposition of nickel oxalate and wave functions of the lithium atom when we were guest research scientists first at Argonne National Laboratory, and later at Uppsala, Sweden, that you were looking for an experienced research director to serve as your deputy. Sam suggested that you might be interested in the details of my credentials for the position.

For the past several years I have been Vice President for Research of a small company of about forty scientists, doing studies for NASA and the Air Force on the attenuation of Electromagnetic radiation by rocket exhaust plasmas and electrical properties of organic crystals. Previously, I had moved to California to the University of California, Berkeley, as professor of chemistry, dividing equally between teaching graduate courses in physical chemistry and research in thermodynamics of real gases. I was invited, at the same time, by Lawrence

Radiation Laboratory and Hughes Aircraft Company, to serve as a consultant. For Hughes, I was asked to review the field of organic semiconductors and to recommend a program of research. Based on my report, a research group was established which is actively and productively at work in this field. For LRL, I provided guidance in physical and inorganic chemistry with particular reference to magnetic phenomena and cryogenics.

Much previously, I had spent a year at the Centre d'Etudes Nucleaires de Saclay, France, where I set up facilities for studies of radiation damage in metals at low temperatures. I met Sam Wortman at the Institute for Quantum Chemistry at the University of Uppsala where we were Visiting Research Scientists in the Quantum Chemistry Group; then I followed him to Argonne National Laboratory as Guest Scientist where we studied the effects of gamma and neutron irradiation on the thermal decomposition of nickel oxalate and clathrate compounds.

Previous to these rewarding experiences, I served as an Advisory Scientist at GE's Knolls Atomic Power Laboratory, where I directed the efforts of a group of chemists, physicists and engineers in the development of reference designs for a 150 Megawatt aqueous homogeneous nuclear reactor.

The attached resumé will provide you with more complete details of these positions and my further background in research, administration, teaching, as well as information on my education. I suppose it is next to treason to live in California and wish to live elsewhere. After six years of trying, I do prefer other environs. My wife and I developed a fondness for the Washington, D. C. area a number of years earlier when I was a visiting lecturer at Johns Hopkins University.

If my background and qualifications are of interest, I'd enjoy exploring further the possibility of filling the position of Deputy Director of Research for Pan American Nuclear Research Corporation. May I hear from you?

Sincerely yours,
Charles S. Najarian

RESUMÉ

Charles S. Najarian
7262 Los Amigos Drive
Rancho Cordovo, California 95670

Fields of Specialization and Interest:

Thermodynamic properties of matter, with emphasis on structural and magnetic transitions; thermodynamics of reactions; cryogenics; inorganic chemistry; reaction kinetics; spectroscopy; electrochemistry

Professional Experience:

1965 to present—Vice President for Research, Advanced Thermodynamics Research Corporation, Berkeley, California, directing approximately forty scientists in contract studies for NASA, the Air Force and Institute for Defense Analyses, chiefly on attenuation of electromagnetic radiation by rocket exhaust plasmas, electrical properties or organic crystals, and neutron spectra measurements.

1961-1965—Professor of Chemistry, University of California, Berkeley, Primary duties equally divided between teaching graduate courses and conducting grant research in various fields of physical chemistry: Chemical thermodynamics, Thermodynamics of real gases, Electrochemistry, Advanced Physical Chemistry.

1961-1965—Consultant, Lawrence Radiation Laboratory, Livermore, California. Consultant in Physical and Inorganic Chemistry, with particular reference to magnetic phenomena and cryogenics.

1961-1965—Hughes Aircraft Company, Research Laboratories, Malibu, California. Hughes asked me to review the field of organic semiconductors and to recommend a program of research in this field. Based on my study, a research group was established which is actively and productively at work.

1960-1961—Guest Scientist, Argonne National Laboratory, Argonne, Illinois. Studied the effects of gamma and neutron irradiation on the thermal decomposition of nickel oxalate and clathrate compounds such as guinol-oxygen and guinol-argon.

1959-1960—Visiting Research Scientist, Quantum Chemistry Group, University of Uppsala, Sweden. This association gave me an opportunity to familiarize myself with the "quantum chemical" calculation methods developed by the Uppsala school. I worked on the development of wave functions for the lithium atom which would include correlation effects.

1958-1959—Research Consultant, Centre d'Etudes Nucleaires de Saclay, France. Here I was invited to set up facilities for studies of radiation damage in metals at low temperatures.

1955-1958—Advisory Scientist, Knolls Atomic Power Laboratory, General Electric Company, Schenectady, N. Y. I had two assignments, the first as Supervisor of Systems Engineering for the National Advanced Reactor, and the second as Manager, Chemistry and Ceramics Section. The work in Systems Engineering involved the direction of the efforts of a group of chemists, physicists, and engineers in the development of reference designs for a 150 megawatt aqueous homogeneous nuclear reactor. These reference designs furnished basic concepts for the mechanical chemical and structural engineering groups. The work in Chemistry and Ceramics was oriented primarily toward the needs of heterogeneous reactors, the Empire Reactor and the Belgian BR-3, both of which are in operation.

1949-1955—Assistant and later Associate Professor of Chemistry, Institute for Atomic Research, University of Wisconsin, Madison and simultaneously as Group Leader in Physical and Inorganic Chemistry in the Institute Research Laboratory. I went to Wisconsin after receiving my Ph.D. In addition to my teaching duties, I was responsible for the conception and installation of a well-equipped laboratory for low temperature work and for research in thermodynamics. Much of the equipment had to be specially designed and fabricated in our own shops. Among the major items so built were two adiabatic heat capacity calorimeters for use down to 15 degrees absolute; one calorimeter for use at liquid helium temperatures; an apparatus for measuring magnetic susceptibilities down to temperatures of liquid hydrogen; and a calorimeter for measuring heats of combustion. I directed the efforts of eight men, several of whom were Ph.D. candidates.

1953-1954—Visiting Lecturer and Research Associate, Johns Hopkins University, Baltimore, Maryland, during the academic year 1953-1954 on leave of absence from Wisconsin. I worked on the fluorescence and absorption spectroscopy of isotopically labelled complex compounds of uranium at liquid helium temperatures.

Military Service:

1942-1946—U.S. Navy Research Laboratory repair and installation of radar equipment 1942-1943. Submarine Service in Pacific 1944-1946. I entered the Navy as an Ensign and was honorably discharged into the Navy Reserve as a Lt. Commander.

Education:

Princeton University, Ph.D. 1944
University of Chicago, M.S. 1940
University of Michigan, B.S. 1939
Institute of Quantum Chemistry, University of Uppsala 1959-1960

Professional Societies:

American Chemical Society
American Physical Society
American Nuclear Society
The Chemical Society (London)
The Faraday Society
Sigma Xi
American Association for the Advancement of Science

Patents:

Emergency Shutdown for Reactors
Neutronic Reactor
Logarithmic Amplifier

Personal Data:

Age: 49
Height: 5' 10"
Weight: 165 lbs.
Health: Excellent
Marital Status: Married, three children
Security Clearance: Secret and Q Clearance

104 Marston Hall
Iowa State University
Ames, Iowa 50010

May 20, 1968

Mr. A. Judson
Manager of Technical Employment
IBM Corporation
590 Madison Avenue
New York, N. Y. 10022

Dear Mr. Judson:

The May issue of <u>American Mathematics Monthly</u> had a feature story on IBM that convinced me that mathematicians employed by IBM have envious opportunities for professional growth and personal advancement. As a graduating mathematical statistician, I believe I could do no better than to start my mathematics career under one of your training programs. Won't you consider my qualifications and background for such an assignment?

At Iowa State University where I am receiving my B.S. in Mathematics in June, I have enjoyed an introduction to the IBM line of data processing and computing machines. Under the guidance of Professor Margaret M. Cohen, a former staff member of your Yorktown Heights Research Center, I have gotten my feet fully wet in problems suited for electronic computers and data processors. This instruction included basic principles of machine logic and systems analysis. I have also, under her guidance, received operational programming experience, specifically in instructional coding, program loops, subroutines, flow charts, checking, testing, and operation.

Since the time I enrolled in Iowa State, I have had to meet my own financial requirements. I've been fortunate in the interesting part-time summer jobs that have enabled me to pay my way. Besides financial assistance these jobs provided acquaintance with and practical experience in electrical power, business, practical mathematics, and apartment managing activities. These jobs have given me a better understanding of people and have built up my self confidence and responsibility.

Between working and going to school, I found time to enjoy several activities associated with campus life. The one I benefitted most from was the Sigma Phi Epsilon Fraternity of which I became comptroller in my junior year. Living with the men of the fraternity gave the the chance to meet and mix with at close range persons of different character. As comptroller, I developed a sense of financial responsibility by being trusted with and handling other people's money. I consider this an experience as valuable as any of my class room experiences.

The knowledge, training and experience I have acquired to this point is, I believe, only a start in my education. I hope to mature it considerably by a training program such as yours. I believe an association with IBM would be mutually beneficial. I have not only my background, education and experience to offer you, but sincere enthusiasms and aspirations for advancing myself and, through my humble contributions, IBM. If my letter and data sheet convince you that my background is of interest to your program, won't you have one of your visiting campus representatives grant me an interview to confirm my letter's impression. I look forward to hearing from you.

Sincerely yours,

Richard K. Bryson

DATA SHEET

Richard K. Bryson
104 Marston Hall
Iowa State University
Ames, Iowa 50010
Telephone: 884-9165

PERSONAL DATA:

Age: 21
Height: 5'9"
Weight: 150 lbs.
Health: Excellent
Marital Status: Single
Military Status: Subject to two-year call
Special Interest: Sports, music, youth work

EDUCATION:

June, 1968, Bachelor of Science Degree in Mathematics, specializing in Statistics, Iowa State University, Ames, Iowa

Major Courses in Mathematics:

Analytical Geometry and Calculus
Differential Equations
Matrix Algebra
Applied Statistics
Inferential Statistics
Statistical Theory in Research
Analysis of Variance and Covariance
Machine Computation and Data Processing
Programming for Digital Computers

June, 1964, Diploma, Ames Senior High School, Ames, Iowa

EXPERIENCE:

Part-time—Academic years of 1967-1968, 1966-1967. Resident Coordinator for Iowa State University's new student housing; trouble shooting for the University Housing Department on complaints and repairs.

Summer, 1967—Iowa State University Housing Department; Laborer in the men and women's resident halls.

Summer, 1966–Public Service Company of Iowa, Des Moines; Student engineer on surveying crew of high voltage lines.

Summer, 1965–Public Service Company of Iowa, Des Moines; Laborer with Electric Service Department doing building maintenance.

Summer, 1964–Colorado State Game and Fish Department, Grand Junction, Colorado; Laborer working with a wildlife manager on a big game sampling problem.

UNIVERSITY ACTIVITIES:

Sigma Phi Epsilon Social Fraternity
    Office Held: Comptroller, 1966-1968

Assocation of the U. S. Army, Student Member
    Office Held: Treasurer, 1967

Iowa State University Freshman Welcome Week Committee, 1967

REFERENCES:

Dr. Margaret M. Cohen
Professor of Mathematics
Iowa State University
Ames, Iowa

Dr. Angus K. Moultrie
Associate Professor of Statistics
Iowa State University
Ames, Iowa

Mr. Jed W. Blake
Head, Iowa State University Housing Department
Ames, Iowa

Mr. S. A. Ganz, Manager
Manufacturing Division
Pacific Metal Works, Inc.
508 S. Portland Street
Seattle, Washington 98108

Dear Mr. Ganz:

Please consider my qualifications for the Principle Industrial Engineer you are seeking for your Division to establish standards, methods and procedures for electrical and mechanical assembly. Briefly, here is what I have to offer:

A B.S. in Mechanical Engineering and an M.S. in Engineering Administration;
Eight years of short run production engineering experience in electrical and in the hard goods industries;
Demonstrated capability to plan and implement manufacturing standards and methods for achieving balance work flow;
Patience, emotional maturity, and a desire for new challenges.

You may be interested to know that my military obligations were served as an industrial engineer for the Air Force. After receiving my M.S.E.A., I was commissioned as a second lieutenant with duties as an industrial engineer and later as supervisory industrial engineer, concerned with plant layout, materials handling systems, methods and procedures and still later taught at the Air Force Technical Institute in Dayton courses on quality control, plant layout, and work simplification. I ended my military duties as a major in the USAF Reserve.

The attached resumé provides fuller details of my experience, education, and background. I believe that as you read my qualifications you will see why I am confident I can offer the type of leadership and service you are looking for in a principal industrial engineer.

I shall appreciate an opportunity to discuss this matter with you in greater detail. Won't you let me know when it would be convenient for you to see me? My phone and address are listed in the resumé.

Sincerely yours,
Charles D. Rodman

RESUMÉ
Charles D. Rodman
3708 17th Avenue, N. E.
Seattle, Washington 98105
Telephone: 206 – 283-5200

PERSONAL DATA:

Age: 36
Height: 6' 1"
Weight: 200 lbs.
Health: Excellent
Marital Status: Married, 2 children
Military Status: Major, USAF Reserve

EXPERIENCE:

1965-present—West Coast Metal Products, Seattle, Washington—Manufacturing Engineer, Supervisory (promoted to this position February, 1967). Established standards, methods, and processes for electrical and mechanical production assembly. Established tooling requirements and approved designs of tools and fixtures. Redesigned components and assemblies and coordinated engineering change orders. Provided make or buy decisions. Designed plant layouts, flow charts, and time elapsed charts. As Manufacturing Engineer, monitored the establishment of work factor time standards. Methodized operations and balanced work flows of assembly. Initiated tool and fixture design requirements. Resolved production floor problems.

1960-1965—Northern Oregon Electrical Products—Standards and Methods Engineer. Established work standards and methods for plant operations, using techniques of time studies, standard data, and Methods Time Measurement (MTM). Prepared layouts, flow charts, and routings of manufacturing operations; estimating; job evaluation studies; analysis studies of high versus low volume production and machine utilization.

1956-1960—U. S. Air Force. Commissioned Second Lieutenant promoted to First Lieutenant, Captain, and then to Major USAF Reserve on honorable separation from duty. Supervisory Industrial Engineer at Wright-Patterson Air Force Base, concerned with plant layout, materials handling systems, methods, and procedures. Later as First Lieutenant and Captain concerned with management procedures, production control procedures, process engineering and work measurement systems. Assigned to Air Force Technical Institute. Prepared and conducted courses on quality control, plant layout, and work simplification.

1953-1954—Hewlett-Packard Company—Product Engineer. Interpreted customer specifications and design requirements for sales quotations. Customer engineering contact. Conducted feasibility studies of new products. Redesigned products for cost reduction and economy.

EDUCATION

B.S.M.E. School of Engineering, University of Washington, 1953
M.S.E.A. School of Engineering, Stanford University, 1956

PROFESSIONAL LICENSE

Registered Professional Engineer, State of Washington, License #100003814

PROFESSIONAL MEMBERSHIPS

American Institute of Industrial Engineers
American Society of Mechnical Engineers

ACKNOWLEDGMENT AND FOLLOW-UP LETTERS

Today, not only organizations but also individuals acknowledge or reply to letters of application, particularly unsolicited letters.[1] Below are typical acknowledgment or reply letters.

(a) Dear Mr. Hall:

Thank you for answering our advertisement in the Chemical & Engineering News for a Radiochemist.

In order that your qualifications may be considered fully, we suggest you complete and return the enclosed application form. Please include all pertinent information so that we will have as complete a record as possible of your experience and education.

---

[1] This practice is not followed in instances where "blind ads"—applications to be sent to box numbers—are placed in classified advertisement sections of newspapers and magazines; the chief purpose of not identifying the advertiser is to avoid the cost involved in replying to applicants whose qualifications are inappropriate for the opening. In some organizations the cost of sending a letter is more than 100 times the cost of the six-cent stamp on the envelope.

When we receive this information, we will carefully review your background and I will contact you again shortly.

We certainly appreciate your interest in our company and look forward to hearing from you.

Sincerely yours,

Donald B. Adams
Employment Manager

(b) Dear Mr. Hall:

Thank you for completing our application form.

We would like you to visit us to discuss possible employment. Please telephone collect area code 301-343-6100, Extension 2228, Mrs. Rice, to arrange a convenient interview time.

Sincerely yours,

Donald B. Adams
Employment Manager

(c) Dear Mr. Peterson:

I am confirming your telephone arrangements. We will be expecting you in the Personnel Office on Tuesday, June 18, 1968 at 9:30 A.M.

We have made a reservation in your name at the Leamington Motor Inn, 7th Street and Roosevelt Boulevard, for the night of Monday, June 27. When checking out, please sign for the room charge so that this may be billed directly to us. We will reimburse you for meals and travel expenses.

We have also made arrangements for you to pick up a prepaid, round-trip airplane ticket at the United Desk, Stapleton Airport.

Your flight schedule will be as follows:

Date	Airline & Flight No.	Time Leave	Time Arrive	From	To
6-27	United 287	5:30 P.M.	10:35 P.M.	Denver	Baltimore
6-28	United 288	6:55 P.M.	8:50 P.M.	Baltimore	Denver

I would appreciate your completing and returning the enclosed application form prior to your visit so that we can use its information as the basis for further discussion about your background.

I shall look forward to seeing you on June 28.

Sincerely,

Clement Parks
Employment Manager

When acknowledgments or replies must be negative, they should be courteous, as follows.

Dear Mr. Franklin:

Thank you for your inquiry about the engineering position we advertised in the New York Times.

We have now had the opportunity to thoroughly review your correspondence. While we are generally impressed with your background, unfortunately, there are other candidates whose specific qualifications are more closely related to our immediate needs.

I am sorry I cannot be more encouraging now. However, we certainly appreciate your interest and wish you success in finding a challenging position.

Sincerely yours,

Fred J. Whipple
Professional Employment

If you should receive a letter like the preceding one and are interested in working for the company, it would be good strategy to reply:

Dear Mr. Whipple:

Thank you for acknowledging my application for the engineering position with your company. Although I regret that there is nothing appropriate for my background at this time, I hope there will be an opportunity for which I may qualify in the near future. I shall look forward to you again examining my resume when such a vacancy occurs.

Sincerely yours,

William Franklin

Your letter serves the purpose of impressing the employer with your interest. Consequently, he may make it a point to tag your application for favorable action at the first opportunity of an opening you might fill, even though it might not be the job you applied for.

The application letter is a sales letter. Good salesmen know that a sale frequently depends upon repeated efforts and follow-ups. Follow-up letters, after an interview, create a favorable impression upon a prospective employer. People admire determination and persistence. After you have received an interview, express your appreciation and reemphasize the particular qualifications that you have determined, during the course of the interview, are important to the prospective employer. If you are able to give additional information or additional qualifications to enhance your application, do this in your follow-up letter. The follow-up letter should be neither too long nor too short, and it should be timed to reach your prospective employer within a week after the interview. An effective example is given below.

Dear Mr. Ganz:

My two hours with you and your enthusiastic group of factory supervisors last Thursday was a gratifying and enjoyable experience. Thank you for acquainting me with your plant, your products and the men with whom the Principal Industrial Engineer you are planning to hire will work.

Since my visit, my thoughts keep returning to the challenges facing the person you will hire to methodize your manufacturing operations and to balance the work flow. Your expansion into the fourth floor of the L-shaped building next door provides an interesting assembly floor problem. As I was shaving this morning, I had a thought of how advantage could be taken of the narrower portion of the loft for the electrical fabrication assembly. I'm attaching some sketches for laying out the work and the floor plan for this idea. This approach could be pilot-tested before the move with a Methods Time Measurement study. If you would like elaboration of the sketches, I'd be happy to provide it.

You can see, Mr. Ganz, I am enthused about the possibility of tackling some of your production problems. Your opening seems to be an opportunity I've trained for in Engineering School, graduate school and for more than eight years as a manufacturing and Standards and Methods Engineer. I am confident I can contribute to the continuing expansion of your company and grow with its success. I would like very much the opportunity to prove this confidence for you. I look forward to this possibility.

Sincerely yours,

Charles D. Rodman

Below is a letter written by an overanxious applicant (it would have been better for him not to write than to put an inept foot in his mouth as he did at several points in the letter):

Dear Mr. Webster:

Thank you very much for extending me the honor of an interview yesterday and telling me about the very demanding administrative assistant position in your research laboratory you are seeking to fill.

I am obviously short of qualifications in some of the areas discussed, but I feel that with some guidance and assistance, these shortages will be made up.

Recognizing the caliber of some of the personnel with whom I would be dealing (should I be the successful candidate) I should like to mention two items which do not appear in my resumé. These items serve only to illustrate my ability to work successfully with individuals outside my normal work.

First, I have been the Executive Secretary for the Philadelphia Contract Bridge Society. This is a local chapter of the American Contract Bridge Society. In this capacity I manage and coordinate all chapter efforts in cooperation with the President and other officers. I am responsible for all chapter records, and for advising and keeping the membership (particularly the officers and the Board of Directors) acquainted with all chapter business. This chapter has seen fit to send me to Kansas City, Washington and Milwaukee to represent the membership at national officer conferences.

Second, the fourteen years I have been a member of the Philadelphia Dental Bowling League—the only non-dentist on it. I have been Secretary and Treasurer for the league during the past eleven years. Association with the bowling league has resulted in close working relationships with many dentists engaged in all specialties of dentistry as well as medical personnel and detail men. (Our league is now accepting MD's in its activities.)

I can supply numerous references concerning these activities should you desire.

In closing, I affirm my active interest in the administrative assistant opening in your research laboratory. I am sure I can contribute to the success and effectiveness of your research organization if you will give me the opportunity and have patience with me at first. I am available for further discussions at your convenience.

Sincerely yours,

THE REFERENCE LETTER

Most employers require an applicant to furnish the names and addresses of two or three individuals who can be referred to for information concerning his qualifications and character. These reference letters are usually brief, specific, and courteous. They simply state the purpose of the letter, request information and, express appreciation. For example:

Dear Dr. Zettlemeyer:

This company is considering Dr. Henry T. Plummer for a technical position. He has given your name as a reference.

We would appreciate it if you could inform us of your association and professional relationship with him, as well as your frank opinion of his strongest area of technical interest and competence: His ability to work with others and his relative rank among persons with similar training and experience would also be of great interest to us.

Any additional information which would help us form a complete evaluation of Dr. Plummer's abilities, growth potential and value to our company would be extremely helpful.

We shall appreciate your cooperation in this matter. Your information will be held in strict confidence. Should we be able to return a similar favor in the future, please call on us.

Very truly yours,
Edward S. Marquis
Employment Manager
Normandy Instrument Co.

THE LETTER OF RECOMMENDATION

Professional persons are often called upon to write letters of recommendation on behalf of an applicant for a position. The purpose of this type of letter is to give a prospective employer pertinent information about the applicant's qualifications, abilities, character, and general conduct. Whether he is a colleague, friend, acquaintance, or employee, the letter should reflect an honest and candid appraisal of the applicant. An applicant should always notify the persons he intends to use as references to be sure, first, that they are agreeable to the task and that they feel qualified to provide the information. (This courtesy of obtaining permission saves the applicant, the prospective reference, and prospective employer much embarrassment.) Second, prior notification allows the person more thought for preparing and adapting his recommendation to the situation. Very few persons will refuse, outright, a request for a recommendation, and many write perfunctory ones. The reader receiving such a recommendation on an applicant does not have to be a shade above a high-class moron to recognize a vague and tepid statement. Therefore, when you are asked to furnish a recommendation, be sure you will not have to resort to "damning with faint praise." I do not intend to imply that every recommendation you write should read like a eulogy at a funeral. Employers are interested in knowing not only the strengths of the applicant but also his weaknesses. Employers will mistrust the recommendation that seems exaggeratedly favorable, as if written by a marriage broker. My suggestion is that you ask yourself what you would want a letter to contain if you were the reader. With this in mind, the letter should be easier to compose; its content will be fair to the applicant and helpful to the reader. Consider the following letter.

> Dear Mr. Balch:
>
> I am replying to your letter of October 2, regarding Mr. Andrew Springer whom you are considering for employment. He worked, as he says, for my department. He was a satisfactory employee. I wish to inform you that I believe you will find him satisfactory, too.
>
> Very truly yours,

The above letter (unlike the unhappy applicant, Mr. Andrew Springer) is highly unsatisfactory. The letter's laconic and exasperating tone gives the reader little to go on. If Springer were and would be satisfactory, the writer should state why—offer specifics in chapter and verse. Otherwise, the reader will search between the lines for inferences that may or may not be there.

Of course, replies often depend on the questions asked in the letter requesting appraisal or evaluation, as did the previously cited letter from the Normandy Instrument Co. Below is a letter requesting appraisal. Although general in its approach, the reply that follows it is competently specific.

Dear Dr. Pardoe:

Dr. Jeremiah M. Benson who is with the National Bureau of Standards at Boulder, Colorado is among those whose names are being considered for the position of Dean of the College of Engineering at State University. It has been suggested to the faculty committee which will make recommendations to the administration that you might be willing to write us about your impressions of Dr. Benson's suitability for such a position.

Our Engineering College is growing rapidly, both in number of undergraduate and graduate students, as well as expanding the area of its research programs. We need as Dean someone who can both give dynamic direction to this growth, and at the same time work harmoniously with department chairmen to upgrade the calibre of their graduate faculties. We will be most grateful for your appraisal of Dr. Benson.

Sincerely yours,

Hiram W. Nageley

Dear Dr. Nageley:

I am glad to tell you what I know about Dr. Jeremiah M. Benson and to offer my appraisal of his suitability for the deanship of your Engineering College, limited, of course, to my knowledge of him.

In my judgement, Dr. Benson would be an asset to any academic institution. I first met Dr. Benson ten years ago when he was a member of the staff of the United Nation's Atomic Energy Commission in Paris, concerned with problems of radio-active waste management on an international scale. At this time I was the Scientific Advisor on the Staff of the American Ambassador to France and had close contact with Dr. Benson's work. His approach to the problem made complex by both technical and political aspects was creative and did much to produce a solution that was feasible and harmonious. He demonstrated in Paris that he had the ability to combine technical considerations with political and administrative judgement to produce a program so effective that it is still, after ten years, the standard operating procedure.

Since he and I have left Paris, I have had only casual contact with Dr. Benson at meetings and at his laboratory at the National Bureau of Standards. I cannot claim to have a specialist's knowledge of the work he has done at NBS, but I have been impressed by his sound judgement and the tangible evidence of his accomplishments. The high regard with which Dr. Benson is held among his professional peers is demonstrated by his being frequently asked to serve as a member of special panels convened by the National Academy of Sciences-National Research Council and by the U. S. Atomic Energy Commission.

Though Dr. Benson has a quiet personality, he inspires confidence immediately. Having seen him in action, I would say he is one of those rare individuals, who can work productively with all his associations. In manner and speech, I would say, he prefers to operate in low-key, but if the situation demands it, he can, with ease, become vigorous and forthright without showing any indications of being overly aggressive.

You may also be interested to know that his wife has been very active in the State Poverty Programs in the Boulder-Denver area. She has also served as a member of the Governor's Committee planning the statewide Colorado medicare program.

In short, I can think of no one whom I would be happier to recommend for an academic position in your university.

Sincerely yours,

For study purposes here is another letter of recommendation:

Dr. Paul H. Cheifetz, Chief
Nuclear Physics Section
Materials Research Division
National Bureau of Standards
Washington, D. C. 20234

Dear Paul:

I understand Mrs. Karen Deutschmann has applied for a position in your section. I am glad to be able to commend her to you. She has been working in my Branch for the past four years. Unfortunately, for us, she will not be able to move to Tennessee when we transfer our activity to the Oak Ridge National Laboratory next month, choosing to stay in the Washington area so she can pursue part time her doctorate degree at the University of Maryland.

Mrs. Deutschmann is an exceptionally intelligent, diligent, and conscientious person. She identifies herself with the project in hand in a very responsible way, with the result that one feels that any job she has taken on is in the hands of someone with push and judgement. She will need a certain amount of advice and guidance in an area with which she is not familiar, but she will always feel that a job has to be as nearly perfect as it is humanly possible to make it and that she is there to help achieve this perfection. As you know, this a quality beyond any price.

She learns very quickly. Although prior to joining our operation she was not specially trained in nuclear physics, she picked up the fundamental requirements for her work in our projects in nuclear decay and reaction properties in a short time and has been a creative contributor.

You are probably already acquainted with the details of her education and experience from your interview and her resumé. To these I would like to add that she is exceptionally friendly, cheerful, and kind and able to work well with men in the laboratory at their own level. In short, she is the sort of person who increases the good working spirit of any group. You will find her an asset to your programs as I have.

Cordially,

Sam H. Pogorny

LETTERS ACCEPTING POSITIONS

It is customary for an applicant—even though acceptance of the job has occurred in a face-to-face meeting—to accept in writing the position that has been offered him. It is not only a courteous and appreciative gesture but it also completes the "contract." Here is an example:

> Dear Mr. Trowbridge:
> I am happy indeed to accept the staff scientist position in your company. I consider it an honor to be selected to join the research laboratory.
> The conditions of my appointment as outlined in your letter are in accordance with our previous discussions and are certainly acceptable. I have tendered my resignation of my present position and shall be able to report to the research laboratory one month from today.
> I am looking forward with enthusiasm and confidence to this new association.
> Sincerely yours,

THE LETTER OF RESIGNATION

The professional employee sometimes finds it necessary to resign from a position, either to accept a better position or because ill health prevents him from continuing in his present job. Out of courtesy, resignations are often given face-to-face. Even so, most organizations, as a matter of record, request that resignations be written or confirmed in writing. The letter should include the reason for resigning, an expression, perhaps, of appreciation or regret (or both), and a definite date when the resignation is to become effective.

Certainly, there are times when unhappy and frustrating circumstances force an employee to sever his relations with an organization and go elsewhere to follow his profession. No matter how badly or unjustly you may feel you were treated, it is better to effect the change gracefully and harmoniously. The resignation letter is the formality of record. Say goodbye graciously. You do it by saying something complimentary. Of course, I do not mean for you to tell a lie. But surely, in every situation, there were opportunities for gaining experience, or there were associates who helped in some way, or who were considerate, or who sharpened your wisdom. Say so in the resignation letter. Here is an example to study:

> Dear Mr. Dibelka:
> A week ago the General Aerospace Corporation of Houston, Texas, invited me to join its organization as Chief Mechanical Designer at a most

attractive salary. After much deliberation and discussion with individuals whose judgement I value, I have decided to accept General Aerospace's offer.

I have been with Eastern Engineering for more than eight years; these years have been productive and pleasant. I have enjoyed growing with the company and will always remember my stimulating association with you and the very fine design engineers of the department. I shall always be grateful for the many challenges which came my way at Eastern Engineering. I know they sharpened my design capabilities and developed my professional competence.

However, the General Aerospace offer will permit me to work in avionics, a field in which, as you know, I have a strong interest. I would be doing an injustice to my family and myself if I were to forgo the opportunity being offered me.

Please consider this letter my resignation. To permit you time to find or train my successor, I request to be released from my duties one month from today—August 1, 1968.

You, the members of the design department, and Eastern Engineering have my best wishes for continued success in your many enterprises.

                                        Cordially,
                                        Clifford C. Maynard

THE LETTER REFUSING A JOB

Fortunate is the applicant who has several jobs to choose from. When you have such a conflict of riches, it is better to advise the organizations whose offers you are refusing as soon as your employment is firm. Tell them that you are no longer interested in the job offered. The letter should express gratitude for the offer and should explain graciously why another offer was accepted. The courteous tone of your letter should induce them to leave their welcome mat out in the event that fate and circumstance, at some future date, reengage your interest in an opening with that company. Here is an example of this kind of letter:

Dear Mr. Allison:

I am very gratified to receive your offer of employment as a junior mycologist in your clinical laboratory. I have discussed it with my major professor at State University, Dr. Tracy Wilkens. He tells me I should be flattered by it and I am. Dr. Wilkens believes that I would find no better opportunity to pursue microbiology than in your laboratory.

Your offer has given me not only a great deal of happiness but conflict also. You see, I had already decided to be an aquatic ecologist before it arrived and had been considering a similar offer from a laboratory in Cambridge, Massachusetts.

After much and further discussions of my career aspirations with both Dr. Wilkens and my parents, I have come to the decision that my interest in aquatic ecology can best be developed further by Ph.D. study. The offer from the other laboratory permits me to continue part time, advanced study at company expense. Harvard University has accepted my application for this study. Under these circumstances, so advantageous to my aspirations, I have decided, not without mixed emotions, to accept the offer from the laboratory at Cambridge, Massachusetts.

I do appreciate the very wonderful opportunity you extended to me. I shall always recall the pleasant experience I had during my interview visit to your laboratories. I know it would have been a joy to have been part of your laboratory organization.

Sincerely yours,

Harold Becker

9
PERSONAL AND NON-TECHNICAL CORRESPONDENCE

All of us occasionally write letters that are of a personal nature, and do not relate directly to our daily work or professional activities. Among this kind of correspondence are letters making hotel reservations, luncheon or dinner appointments, letters of introduction, recommendation, congratulations, condolences, and other similar types. Because the technical professional's personal letters usually pertain to less formal situations than those related to his work, they are less ceremonial. They have an air of ease and casualness. They reflect the personality of the writer more than does his business correspondence. Individuality plays a strong role in the letters. So do the characteristics of sincerity and courtesy. Style and tone in personal letters depend largely on the degree of intimacy between the writer and the receiver. On the whole, the style follows the approach of the business letter.

Format considerations and placement of elements of the letter correspond to those of the business letter. If the recipient is intimately known (on a first-name basis), then the salutation utilizes the first name in its greeting as, Dear Fred:, Dear Tom:, Dear Margaret:. This applies also to the business letter. In the personal letter, if the recipient is not intimately known, then the formal salutation such as, Dear Dr. Brady:, Dear Mrs. Weisman:, Gentlemen:, or Dear Sir: is used.

In correspondence between friends or business associates of longstanding (even in the conduct of professional and business affairs), the use of an intimate patois, colloquialisms, and humor is not only permissible but desirable. Conciseness, however, is important because it aids clarity and saves the reader time. Clarity and succinctness are desirable virtues in all writings of a professional.

LETTERS OF RESERVATION

Professionals are always on the go—business trips, professional conventions, committee meetings, symposia. One of the most frequent types of letters is the one making a hotel or motel reservation. Although substantively brief, the letter must be clear, specific, and detailed about requirements. For instance:

Gentlemen:

Please reserve for me and my wife an outside room with double bed and shower for February 3-7 inclusive. I am recovering from a chipped bone in my foot, so I would appreciate a first floor room, convenient to your dining room and parking facilities. I understand you offer reduced rates to university and college faculty. I am on the faculty of State University and would appreciate your extending me this courtesy. Please confirm this reservation and the rate of the room.

Sincerely yours,

LETTERS OF INTRODUCTION

The professional is often called upon to grant personal favors. Among the most common obligations is to write a letter of introduction. The letter helps to open doors that might otherwise be closed. If the person being introduced is well known to the writer and has his confidence, the letter should indicate this. If the person being introduced is only casually known, the writer has a responsibility to the recipient to indicate unmistakably the extent of the relationship. The organizational construction of the letter is simple; its language is concrete and clear; and its tone is considerate and courteous. It's elements are: (1) a statement of introduction; (2) a descriptive identification of the bearer of the letter; (3) a statement telling why the introduction is being made; (4) a specific request to the recipient to provide the requested courtesy; and (5) an expression of appreciation for the consideration of the request. Here is a letter making good use of these points.

Dear Steve:

The purpose of this letter is to introduce Peggy Standish as a possible job candidate for your rapidly growing organization.

For approximately three years, Peggy acted as my good right arm. Her basic responsibilities were: (1) handling administrative matters connected with the chemical, biological, and analytical data program; (2) coordinating our literature, advertising and exhibit activities; (3) performing secretarial work for me.

Although Peggy has had no formal training in administrative work, nor in advertising and technical writing, she was able to develop these skills to a very competent degree through actual work experience.

She is not only a pleasant person, but also has initiative, is hardworking, loyal, and she has that rare ability to think for herself. In summary, I believe her experience and potentials in administration, writing and sales in the fields of chemistry, biology and analytical data could prove to be a valuable asset to your organization.

She took maternity leave of us more than five years ago. Her youngest is now in school and Peggy wants and needs a job that can exploit her capabilities. Unfortunately for us, our backlog of orders is not what it used to be, so we cannot take advantage of her availability. I believe you would be happy with her work and, just as important, Peggy would enjoy contributing toward making your growing organization move further along.

I would appreciate your giving Mrs. Peggy Standish a few moments of your time. I am confident you will be glad to make her acquaintance.

Cordially,

Here is an example of a letter of introduction for an individual with whom the writer has only a casual acquaintance.

Dear Ed:

The bearer of this letter is the son of a very old and dear friend of mine, James L. Banks, of whom you have heard me speak a great deal. Young Jim Banks, Jr. has just finished college. While waiting for Uncle Sam to summon him into the Navy this next fall, he is doing some interim work of a survey nature for a firm in Philadelphia. It has to do with polling executives on how bearish or bullish they are about their own company's prospects and business in general. I'm sure Jim can explain the details much better than I.

I would consider it a great favor to me if you gave the young man about an hour of your time and opened the door for him to meet your two partners as well. He seems a bright and pleasant chap. He may eat up a little of your time but I don't think he'll bore you—and you might find the survey he's doing interesting in its own right.

Both Jim Banks, Sr. and I will be grateful for any courtesy you show the young man.

And let's get together for golf soon!

Best to the frau,
Terry Smith

LETTERS OF INVITATION AND LETTERS OF ACCEPTANCE OR DECLINATION

The technical professional has many occasions to send out letters of invitation. He may invite a person of prominence to speak at a professional meeting, to a gathering of associates, or to participate in a panel. On a more personal and social level, he may invite a friend to have lunch or dinner. Invitations to speak should include: a background explanation of the occasion; a statement indicating why the recipient was chosen; designation of the subject and length of time of the speech; an indication of a suggested approach, if that is appropriate; other details of the program, if any; and composition of the audience. Invitations to attend a meeting or function need to include a brief description of the purpose of the meeting, a statement of the time and place, and an expression of the writer's desire for the recipient to be present or a statement indicating to the reader the meeting's value.

Invitation to Participate in a Committee

Dear Dr. Del Plaine:

The Program Committee for the 1969 American Geological Society annual convention scheduled for May in Dallas would like to have an area representative from the Rocky Mountain Section. I am writing you because Dr. Gunther Muelhouser, President of the Society, has enthusiastically recommended you for this post.

The committee is now being formed and will consist of members from approximately twelve local chapters to provide full geographical representation. The task of each committee member is to help plan the convention program, encourage the submission of good, provocative papers from various disciplines, procure speakers and panelists, follow-up your self-generated and other contacts, screen papers, and help, if necessary, in preparing the convention program brochure.

To accommodate the widespread geographical distribution of the committee, two planning meetings will be held. One has been arranged for Friday, October 15, in New York City. The other will be, I hope, in Denver on or about Friday, November 5.

Dr. Muelhouser and I hope you are interested in taking on this important assignment. As area representative from the Rocky Mountain Section, you might guess that we will ask you to take on the additional assignment of setting up a meeting place for November 5.

The Society and I are looking forward to receiving your acceptance.

Sincerely yours,

An Invitation to Speak

Dear Dr. Louviers:

The American Institute of Physics has asked me to organize a full day's symposium on problems of information transfer among scientists. The emphasis in this symposium is to be on the role a government-wide, coordinated national scientific and technical information system might play in helping to meet these problems.

This symposium would be incomplete without a presentation to the audience of a broad view of the range of thinking and planning now underway within the non-government component of the technical community in the area of scientific communication. In your capacity as Chairman of the National Academy of Sciences—National Academy of Engineering Committee on Scientific and Technical Communication you are undoubtedly the best qualified person in the United States to speak on this subject. I am writing, therefore, to inquire whether it might be possible for you to participate in the planned symposium.

Most of the talks are being scheduled for thirty minutes, including discussion, but I consider that any talk that you might give would be of such great interest that any amount of time that you might wish to take would be in order.

The symposium has been set for January 28; I hope that date is convenient for you. A copy of the tentative program is attached. Most of the listed speakers have already accepted their assignments.

I would appreciate your letting me know at your earliest convenience whether you will be able to participate, in order that the plans for the program can be completed within the deadline that I have been given.

Sincerely yours,

Professionals often have occasion to combine business with pleasure. Friends are usually involved in similar interests and activities. Invitations to friends are informal and include references to personal matters. These letters follow a conversational and unceremonious tone.

Dear Ed:

By now, world traveler, you should be back to the unexciting, mundane existence of running a technical publishing company. After a month in Europe and other exotic places, even Colorado should be dull. Once you get your desk cleared off, perhaps you'll be itching to travel again and might meander in the direction of Washington, D. C.

Seriously, it's been almost a year—far too long—since we've gotten together. I'd like to know how your new publication effort is progressing. Should your travels send you out this way, please give me a call. My home telephone is area code 301, 395-3662. My office is area code 301, 921-2228. Incidentally, my activity has recently moved to Gaithersburg, Md., but is no more than half an hour from the center of Washington.

I am attaching a descriptive brochure of our program. The readers of your various technical magazines should be interested in our story. Should you wish to do a piece on our program, I'd enjoy providing you with details.

Perhaps serendipity would take place in our getting together!—I'm suggesting each of us an excuse other than friendship for an early meeting.

We've had Colorado weather here into December—the warm, balmy Indian summer kind. And frankly, I don't miss the snow which, I gather, Santa brought out your way. So leave your skis at home and come share the sun!

Cordially,

Letters of Acceptance or Declination

Courtesy requires the recipient of an invitation to reply promptly. Letters accepting or refusing the invitation are composed in the same tone of formality or informality as the invitation. Here are some examples:

Dear Dr. Gratz:

I deeply appreciate your gracious invitation to speak at the American Institute of Physics symposium being held in New York City January 28. It is an honor I am glad to accept.

It is my understanding you wish me to discuss the National Academy of Sciences-National Academy of Engineering considerations for the evolvement of a national communication network for the dissemination of scientific and technological advances. This I shall be happy to do. Since I have always been a firm believer that even the best of speakers can say all there is to be said about a subject in thirty minutes, I wish no more time than that. I shall be pleased to respond to questions after that.

I am looking forward to seeing you and participating in the symposium.

Sincerely yours,

Dear Dr. Scoville:

I shall be glad to serve on the Program Committee for the 1969 American Geological Society Convention as the Rocky Mountain Section representative. I have discussed your invitation with the man I report to in my company and, I am happy to say, have received his warm approval to travel and to take as much time as is needed for work on this committee.

Because desirable meeting space could be a problem at the indicated time of the year, I have already contacted the Brown Palace Hotel and have secured a meeting room and assurances for accommodations for twelve people for November 5.

I am looking forward to working with you and the other committee members toward constructing and implementing a very productive convention program.

Sincerely yours,

Letter of Declination of Appointment

Dear Dr. Rohner:

Thank you for your letter of February 17 inviting me to become a member of the Washington Academy. I have given this matter my most serious consideration and after much soul searching have come to the conclusion that it would be best for me not to accept. Let me explain.

In the present information explosion, our only salvation lies in narrowing, rather than broadening, our fields of interest. I belong to a number of technical societies; it is easy to be tempted into joining another one—especially as eminent as the Academy. I recognize the clear, tangible benefits I would derive; yet, at the same time, I can not but feel such a move would infringe on my present heavy commitments to several other societies.

I cherish the recollection of days when the Washington Academy did so much to nurture the development of the Gnotobiotics Association of America, and all the assistance I received from you in the childhood years of the <u>Gnotobiotics Society Journal</u>. This only makes it more painful to answer your gracious invitation with a negative decision, but I am convinced that it is the proper one.

Sincerely yours,

Letters of Appreciation and Congratulation

One of the major purposes of letter writing is to promote good will and to cement friendly relations. There are certain situations that specifically call for the writing of a letter that will effect better relations between interacting individuals. The professional has many opportunities to write a letter of appreciation for a service rendered, a favor granted, assistance given, or a job well done. The letter of congratulations expresses the writer's sincere interest in someone's success or good fortune. These letters must be sincere because most people easily see through flattery. A hypocritical or pedestrianly phrased letter accomplishes the opposite purpose from the one intended. If you cannot be sincere and enthusiastic in your appreciation or congratulations, it is better not to write. Here are some effective examples.

Dear Dr. Sammartino:

I would like to express my appreciation for the assistance which I recently received from Dr. Tobias Zahn of your staff. I had asked him to referee a controversial manuscript submitted to our Journal. Critical review of manuscripts is a frequent occurrence, but I have rarely obtained such competent, considerate and prompt response. I knew that Dr. Zahn had been under heavy pressure to complete an urgent assignment, and hesitated to ask for his help at all, but because of the controversial judgments generated by this paper I wished some one of Dr. Zahn's specialty and competence to review the manuscript. He accepted this task graciously, saying he had just finished a phase of his major work and needed some distraction to mature his thinking before proceeding to the next phase.

I had his comments in 48 hours. He provided the critical judgments which have given me confidence to publish the paper. His opinions were clearly and succinctly expressed with full and expert substantiation. In addition, he provided very helpful editorial assistance. His review had none of the vague or useless suggestions often found in superficial evaluation. His suggested changes ranged from items to improve expression and grammar to some fundamental questions of mathematics, such as some of the consequences of Goedel's proof. I admired his insistence on obtaining complete clarification of questions, the answers to which would strengthen the paper so that it deserved publication.

The present is not an isolated incident but is one of a long series of instances in which the Journal and I have benefitted from Dr. Zahn's assistance. I believe you would be interested to know how much the Society and I have appreciated his ungrudging and able help. He exemplifies the finest in the scientific tradition.

Sincerely yours,

Professor Herbert A. Redpath
Head, Dept. of Mechanical Engineering
Colorado State University
Fort Collins, Colorado 80521

Dear Herb:

Notice of your retirement has only just reached me. Duties in Washington will prevent me from joining your many friends and associates on June 2nd to express my own warmest regards, well wishes and appreciation for the many kindnesses, aid and counsel you so readily proffered me.

I want you, by this letter, to know that I consider you one of the most dedicated and productive educators I have had the privilege of knowing. I am glad, also, that I have been able to count you as a close friend who always gave me unstintingly of his time and wisdom whenever I sought advice on the educational matters that brought us together.

Colorado State University will miss your leadership. You leave behind many accomplishments in engineering education at the university, in the state of Colorado and nationally to serve as an inspiration for those who remain to carry your work forward.

Let me say again how much I regret to be unable to be with you on June 2, to shake you warmly by the hand, tell you personally of my feelings for you, and to wish you well in your new undertakings. I hope happy circumstances will cross our paths again soon.

Cordially,

Dear Dr. Lapata:

Professor Fried and I would like to thank you for the excellent job you did at the recent symposium on Inorganic Ion Exchange Separations at the Pittsburgh American Chemical Society Meeting. We know that both your presentation, "Ion Exchange as a Tool in Inorganic and Analytical Chemistry," and your participation in the discussion that followed helped make the session a success. It is appropriate that you should share this credit with the other symposium speakers.

The lively discussion during the session was evidence of the interest of the attendees at the session on Inorganic Ion Exchange Separations. A number of comments made to me and to Professor Fried afterward reflected the appreciation of the audience not only for the effective presentations but also for the content of the talks. They apparently provided the audience with the type of information they had hoped to gain from the symposium.

Thank you again for your efforts and cooperation in helping us make this symposium so interesting and valuable to the audience.

Sincerely,

160 Applying Correspondence Principles

Dear Dr. Chen:

I have received a number of favorable comments that lead me to believe that our panel on Microbial Toxins at the AAAS annual meeting contributed something important to the field. Personally, I appreciate the trouble you took to report your research and to present it so effectively. I hope that you in turn received some benefit from participating.

Phil Abelson, editor of Science, has suggested that we write up a summary of the panel proceedings for the Meetings section of Science. I, in turn, suggested to Dr. Carl Kimball who served as reviewer for our panel to summarize from the speakers' papers and from the discussion that followed. Would you be willing to be quoted and extracted, provided that you saw a draft of Carl's summary? I hope you agree that this is the kind of thing that should appear in Science.

Sincerely,

Dear Dr. Postwick:

It was with a great deal of pleasure that I read in the current issue of Electronic News of your election to the presidency of the Institute of Electrical and Electronic Engineers.

Please accept my sincere congratulations and best wishes! I know you will give this added responsibility your usual distinguished service.

Sincerely yours,

Dear Jack:

I was delighted to read the story in yesterday's Denver Post about the patent you received on a fuel cell that is an important component in space explorations. Congratulations! I feel almost as proud of my college room-mate as do your lovely wife and doting mother.

Give them Helen's and my warmest regards, and tell them we share their pleasure in your accomplishment. Good luck and many more successes!

Cordially,

Acknowledgments of Letters of Appreciation and Congratulations

It is not only an act of courtesy and good form to reply to these types of letters but it is also a matter of good human relations. The acknowledgment need be only a brief note of gracious acceptance of the friendly expression. When the writer and the recipient are friends or associates of long standing, the acknowledgment may include an expression of affection or regard of the recipient for the writer. Here are some examples.

Dear Mel:

Your gracious and kind letter of July 23 is heartwarming.

In turn, I wish to thank you for all your good offices and highly effective support during my participation in your program. I had a rich and enjoyable experience, filled with all sorts of fruitful contacts and stimulating talk, formal and informal—not the least being the renewed opportunity to see you and your good lady.

All good wishes to you both for the future.

Sincerely yours,

Dear Mr. Willard:

Thank you very kindly for your congratulations and good wishes in connection with my assuming the presidency of the IEEE.

I hope I can emulate to some degree the distinguished services of so many of my predecessors.

Sincerely yours,

Dear Dr. Bartlett:

I am gratified to learn that our panel stimulated interest and favorable comment. Our panel's success owes much to your careful planning and very accomplished chairmanship.

I am quite agreeable to your suggested method of reporting the meeting's proceedings in Science and will be glad to check my statements as recorded by Dr. Carl Kimball prior to publication.

Yours very truly,

Dear Professor Galperim:

You were thoughtful and generous to send me your good wishes in connection with my recent promotion. Your kind letter has added to my happiness.

If you and Mrs. Galperim should ever come out this way, Abbi and I would have a lot of pleasure in seeing you and renewing an old friendship.

Many thanks and every good wish.

Sincerely,

Letters of Condolence

When a death occurs to an associate, an employee, or a member of their family, a letter of condolence is in order. These letters should be simple, sincere, restrained, and in good taste.

Dear Mrs. Reese:

I was saddened to learn of the passing of your husband, Dr. Laurence Reese. I am among many associates at the Research Institute who admired and respected him not only as a very capable biologist and productive researcher, but also as a warm, friendly human being.

I share your loss and sympathize with you. I too shall miss him.

Very sincerely yours,

10
PROFESSIONAL LETTERS AND MEMORANDA

Exchanging information with colleagues about research is among the oldest traditions in science. The origin of scientific journals was the exchange of correspondence between scientists at distant places. Recipients circulated the letters among their proximate co-workers. A letter from one scientist was often routed twenty times over. With the invention and the spread of printing, learned journals began to compete with the epistle as a means for information exchange among scientists. Transformation of the letter to the modern scientific paper did not fully take place until about a century ago. Scientists and their research activities increased; advances in printing technology brought easier and faster publication; research activities could receive greater—if not mass—circulation fairly quickly.

Today, the information explosion has so expanded and proliferated the number of books, journals, and papers being published that no single individual can keep up with his reading. Many leading scientists have again turned to the letter form to exchange information with their colleagues at distant places and to keep abreast of their activities, discoveries, and ideas. The term "silent" (or "invisible") colleges has been coined for this informal exchange. I do not mean to imply that the letter form as a medium of exchange of information fell into disuse prior to the "information explosion." It never lost its utility (as the exchanges of correspondence between Ernest Lawrence and Ernest Rutherford quoted below will show), but the importance of this medium as an effective and direct communication instrument has been highlighted by the exponential increase of printed scientific literature.

CORRESPONDENCE BETWEEN PROFESSIONALS

The purpose of letters between professionals is to exchange information, ideas, and results. An important element in this objective is not only to provide information to a peer but to have it tested and to obtain feedback and stimulation. Although the letter may be formal in tone, displaying the cold objectivity of science, the medium is informal, reflecting work or thought (tentative and in process) in need of critical reaction and comment. The extent of the informality of language varies with the relationship between the writer and recipient. Social amenities are not neglected, and often items of a personal nature are interlaced with the more serious information. The letters between Ernest Rutherford and Ernest Lawrence are not only of historical interest but also exhibit fine examples of information exchanges between two pioneers in nuclear physics.

Lawrence Wrote to Rutherford on 24 November 1936[1]

"I had intended writing you some time ago regarding Dr. R. (Ryokichi) Sagane, who has been with us the past year and desired to spend this year in the Cavendish Laboratory. I am afraid that he has arrived, and therefore words in his behalf now are a bit late. However, I should like to say that we liked Sagane very much; he proved to be a self-reliant and competent experimenter and a congenial personality. I do hope that you will find him an agreeable person to have as a visitor in the Laboratory, for I know that he is very anxious to be with you and will profit a great deal by such a sojourn.

"All of us here are very busy with a number of things. In addition to the nuclear work, we are devoting a lot of attention to biological problems, as I feel that there is important work to be done in this direction as well as in nuclear physics. We are supplying various artificial radioactive substances to the chemists for investigations of chemical problems and to biologists, particularly physiologists, for use as tracers in biological processes. I do hope that in this way we shall be able to contribute to the elucidation of some biological questions. We are also investigating quite extensively the biological effects produced by neutrons. I think we can say pretty definitely now that neutrons do not parallel x-rays in their biological action. Studies of the comparative effects of x-rays and neutrons will doubtless shed light on the mechanics whereby ionization produces effects in biological systems, and of course also there are the possibilities of effective medical therapy with neutrons.

"In some preliminary experiments on a mouse sarcoma, we got indications that neutrons had a greater selective action in killing this tumor than x-rays. Under separate cover I am sending you a reprint of this work. This fall, similar experiments have been carried out upon a mouse mammary carcinoma with similar indications. In these more recent experiments, many more tumors and mice were irradiated with neutrons and x-rays than in the first experiments on the sarcoma, and the new data also indicate a greater selective action of the

---
[1] These letters are reprinted from the article, "The Two Ernests," *Physics Today*, October, 1966.

neutrons or tumor tissue. It seems to me quite probable that neutrons will prove to be valuable in the treatment of cancer.

"We are this year undertaking the establishment of a new laboratory, which might be called a laboratory of medical physics. The organization and planning of the new laboratory is taking a good share of my time this year, but of course I am glad to do it, although I regret I cannot spend full days in the laboratory. Friends of the University have given funds for a new building and equipment, and I hope that by late next fall, experimental work in the new building will get under way. The architects have practically finished the building plans and we are engaged in designing the new cyclotron. Many of us are having pleasure in planning the new apparatus; although doubtless we are deluding ourselves into thinking that the new outfit will be all that a good cyclotron should be.

"For certain experiments in progress we recently further modified our present cyclotron to bring the beam entirely out of the magnetic field, and we are finding the new arrangement one of great convenience for many experiments. I am enclosing a photograph of six microamperes of six million volt deuterons emerging into the air through a platinum window at the end of a tube six feet long. The beam is quite parallel and can be brought out considerably farther if so desired without undue loss of intensity.

"I have heard from several sources that you are very well and very busy—and in view of the latter, I can hardly expect a letter from you, although, needless to say, I should be greatly delighted if you should find time to write a few lines.

"Professor and Mrs. Bohr are coming to Berkeley in March and we all are looking forward to their visit. I wish it were possible to persuade you to visit America also."

Rutherford Replied with Characteristic Enthusiasm for Lawrence's Success

"I got your letter a few days ago, and was very interested to hear of your latest developments in getting a beam of fast particles well outside the chamber. I congratulate you on your success in this difficult task, and I gather you are hopeful to get even stronger beams in this way. The photograph you have sent me is a beautiful one, and I would be very grateful if you would allow me to reproduce it in a lecture I am just publishing called 'Modern Alchemy,' which is an expansion of the Sidgwick Memorial Lecture I gave in Cambridge a few weeks ago. Unless I hear from you to the contrary, I will assume that you agree to this.

"Dr. Sagane visited us this term and he then decided to go for a short tour to Germany and Copenhagen, and is returning here in the New Year to begin some work. He seems a pleasant fellow, but he writes to me that he is finding a difficulty in seeing some of the German Laboratories, as it is necessary to get a special permit from the Government to do so. This state of affairs in Nazi-land is rather amusing, and when some of our men from the Cavendish wished to visit Berlin to see Debye's laboratory, he wrote to Cockcroft that official permission would have to be granted to the Government before he could admit them!

"As to our own work, we are going ahead as usual. The new High Tension Laboratory is nearly completed and we hope to get a D.C. potential of 2 million volts going. We are also making arrangements to run one of your cyclotrons in due course.

"We celebrated J. J. Thomson's 80th birthday on December 18th by giving him a dinner and presentation in Trinity and also an address with signatures from many of the Cavendish people. He is still very alert intellectually, and he was much moved by our little homely address.

"I wish you good luck in the development of your new laboratory and success in your experiments."

It Was on February 11, 1937 that Lawrence Wrote Again to Rutherford

"I greatly appreciate your very interesting letter received some time ago. I know that you are extremely busy and it is very kind of you to write at such length.

"Your account of the state of affairs in Germany is almost unbelievable. One would think with such a scientific tradition the German people could not adopt such an absurd course of action in scientific affairs.

"The dinner to J. J. Thomson must have been a very nice occasion. It is certainly fine that he has such vigor at his ripe old age.

"I am glad to hear that your new high tension laboratory is coming along nicely and that you are also constructing a cyclotron. As I have written Cockcroft, if we can be of assistance in any way we should be only too glad. I have just heard that he is coming over for some lectures at Harvard and I have written him a letter inviting him to come out to see us. I do hope it will be possible for him to do so. I think it is possible that he might be saved some unnecessary beginning troubles by spending a few days in our laboratory operating our cyclotron. Also in a month or so we shall have our new cyclotron chamber for the present magnet practically completed in the shop. This new outfit has quite a few improvements which Cockcroft would probably want to consider in his design.

"During the past few weeks we have been bombarding with 11 million volt alpha particles, studying the radioactivities produced. In addition to those already reported we have been finding many new activities, especially on up the periodic table. Also we have been making some absorption measurements of the radiation from the cyclotron and find that there is a very penetrating component. We do not know what it is yet, but the indications are that the penetrating radiation consists simply of very energetic neutrons. A 7 inch thickness of lead does not cut it to half. According to Oppenheimer theoretical considerations indicate that the mean free paths of neutrons vary as their energy. Hence it may be that the 14 MV neutrons from Be + MV D^2 have mean free paths of more than 50 cms—something like the penetration of the radiation observed. We are continuing with the experiments with the endeavor to get the experimental facts as clear-cut and definite as possible, and I am sure when this is done we shall understand what is going on. Under separate cover I am sending you several reprints."

LETTERS TO THE EDITOR

The professionally conscious technical man is a prolific writer-to-the-editor. His professional and technical journals are excellent sounding boards for airing and exchanging ideas and information. There are two major categories of letters to the editor: "The Communications to the Editor" and the "Editor's Mail Column."

The Communications to the Editor

This is a letter form that permits quick publication and dissemination of new information and results in the format of short summaries of significant research on topics of high current interest. Journals require that the letters record original work and material not previously published in the open literature. Some research is never published in any other form. The letter provides succinct details of the problem investigated, techniques used, data that resulted, and conclusions arrived at. The letters concentrate on results and conclusions and include only a minimum of supporting material necessary for the proper understanding of the significance of the communication. Thus, the published letter serves as a direct and prompt medium for reaching the group of interested investigators. The letter is written concisely to contain only that information which other investigators may require to carry on similar work. English is the preferred language in American and British journals. However, letters in Russian, French, German or Italian will be published, but the title and synopsis in English must accompany the letter.

Conventions of the format of the "Communications to the Editor" vary from publication to publication. The *Journal of the American Chemical Society* publishes letters on research under the sectional title of "Communications to the Editor"; the *Proceedings of the American Mathematical Society* in a section entitled, "Shorter Notes"; *Journal of Colloid and Interface Science* under a section entitled, "Letters to the Editor"; *Journal of Applied Physics* as "Communications"; *Proceedings of the IEEE* in a section called "Proceedings Letters"; the *Journal of Chemical Physics* has a section, "Letters to the Editor," which is subdivided into four categories entitled: "Communications," "Notes," "Comments," and "Errata"; *Journal of Geophysical Research* under a section called "Letters"; *Journal of Biological Chemistry* as "Preliminary Communications"; *Science* titles its section, "Reports"; the American Institute of Physics, which publishes nine periodicals, receives communications in such great numbers that it also publishes a separate semimonthly periodical entitled *Applied Physics Letters;* the American Physical Society has found it necessary to publish a substantial weekly periodical *Physical Review Letters*; the British Institution of Electrical Engineers also has found it necessary to supplement its fourteen periodicals with a separate publication of letters entitled, *Electronic Letters.*

In their published form, these short communications to the editor bear little resemblance to a conventional letter. Although dated and signed by the author of the communication, they are actually short reports. The size of the communication varies from about 200 words to almost 2000 words. Some journals (as, for example, *Journal of Chemical Physics*) charge writers for publication. The organizational structure of the letter consists of an introductory section (which explains the problem and the method of research); the body (which discusses the resultant data); and the terminal section (which offers generalizations or conclusions and, if appropriate, recommendations). Some of these letters contain descriptive sectional headings, an abstract, and references; many have illustrations, tables, and charts to clarify significant points.

Below is an example of a typical letter appearing in the September 15, 1966 issue of *Physics Letters.*

THE ABSOLUTE THERMOELECTRIC POWER OF LIQUID METALS

A. S. MARWAHA
Birbeck College, London

and

N. E. CUSACK
University of East Anglia

Received 28 July 1966

> The results of new measurements of the absolute thermoelectric power of ten liquid metals are reported. These measurements extend up to $1000°K$ for most metals and include Al and Tl, for which there seems to be no other published data.

This letter presents results of new measurements of the absolute thermoelectric power of ten liquid metals. The most recent comparable values are those of Bradley [1] and the present work is by a technique different from Bradley's, covers a somewhat wider temperature range and adds values for the thermoelectric powers of liquid Al and Tl which are the first available as far as we know.

The basic method was essentially that of Cusack, et al. [2] with some technical improvements. The Seebeck voltage of a thermocouple consisting of the liquid metal and a pure copper or platinum wire was measured and fitted by least squares to a parabola; this could always be done within experimental error leaving the experimental values distributed randomly about the parabola. The thermoelectric power was obtained by differentiation with respect to temperature and the absolute thermoelectric power of Cu or Pt was subtracted. The value used for Cu was:

$$S_{Cu} = 0.05 + (5.45 \times 10^{-3}) \, T \, \mu V/deg.$$

For Pt, values given by Cusack and Kendall [3] were used.

In all cases S could be represented by an empirical formula

$$S = a + bT \, \mu V/deg$$

where T is the absolute temperature. The values obtained were as given in table 1.

Table 1

Metal	$S(T_m)$ ($\mu V/deg$)	a ($\mu V/deg$)	$10^3 b$ ($\mu V/deg^2$)	Temperature range (°K) T_m to:
Al	-2.1	0	-2.24	1250
Bi	-0.7	0	-1.31	1000
Cd	0.5	0	0.79	1000
Ga	-0.4	0	-1.28	1000
Hg	-3.5	2.00	-23.33	585
In	-1.0	0	-2.38	1000
Pb	-3.4	0	-5.72	1000
Sn	-0.5	0	-1.03	1000
Tl	-0.5	1.42	-3.38	1000
Zn	0.1	-2.94	4.42	1000

The error in S is difficult to estimate but the absolute value should not be in error by more than ± 0.2 $\mu V/deg$.

It is possible to calculate a theoretical value for S from Ziman's theory as is done by, for example, Sundstrom [4]. For this purpose it is necessary to choose from the literature pseudopotential matrix elements, $U(K)$, and structural information, often represented by the structure factor $a(K)$. There is by now quite a variety of $U(K)$ and $a(K)$ to choose from (see for example Wiser [5] and so sensitive to the choice is the calculated value of S that we consider it impossible at present to test the theory itself by comparing the experimental and calculated S. For example different choices of $U(K)$ and $a(K)$ give calculated values of $S_{Pb}(T_m)$ equal to -4.4, -0.03, 0.2 and 1.1 $\mu V/deg$.

This latter point and further technical details of the experiment will be discussed more fully elsewhere.

1. C. C. Bradley, Phil. Mag. 7 (1962) 1337.
2. N. E. Cusack, P. W. Kendall and A. S. Marwaha, Phil. Mag. 7 (1962) 1745.
3. N. E. Cusack and P. W. Kendall, Proc. Phys. Soc. 72 (1958) 898.
4. L. J. Sundstrom, Phil. Mag. 11 (1965) 657.
5. N. Wiser, Phys. Rev. 143 (1966) 393.

THE EDITOR'S MAIL COLUMN

This feature in the professional and technical journals, similar to the editor's mail column in the daily newspaper, is very popular with readers. Even the most introverted scientific professional, given an instrument to "sound off," cannot resist the "soap box" from which he can reach his peers. He cannot resist the opportunity to become a pundit, arbiter, critic, commentator, pedantic grammarian, or prolocutor. And this is as it should be. The measure of the professional is his capability and willingness to stand up to be heard on the matters and issues that affect him, his scientific field of interest, his profession, and his conscience. He is eager to share, exchange, and expose his thoughts, ideas, and opinions with his compeers. This freedom of exchange and dialectic discussion is a cornerstone of the scientific tradition. That is why the editor's mail column is lively and readable in both the venerable professional journal and the pedestrian technical or trade magazine.

As in the daily newspaper, letters to the editor in the technical journal are stimulated—often provoked—by items or articles appearing in the pages. Writers will be quick to point out errors in spelling, errors of fact, theory, or opinion. Rebuttals by the original writer or by those of his persuasion quickly follow criticism. Frequently (and, often, as a matter of editorial policy) the writer whose piece is criticized is given an opportunity to reply or to defend his position immediately following the letter of attack. Controversies may be continued by letter through many issues until they have run their course or until a new and more topical one has squeezed the older one from the limited space available.

Some journals limit the amount of words per published letter; others may not. Unlike the daily newspaper whose letter writers frequently request the omission of their name from the published letter, the convention in the professional and scientific journals is that the writer must sign his name. However, letter writers will often request that their organizational designation be omitted to prevent any assumption that the writer's opinion reflects the opinion of the organization for which he works.

Letters to the editor, in their published form, follow the style requirements of the journal to which they are submitted, but the business format is appropriate and should be used. Do not use your organization's stationery unless you particularly want your views identified with your organization and have assurances from your management that it does not object to this identification. Type your letter. No editor will take time to decipher your scrawl. Do not be abusive. Editors are wary of letters that could be interpreted as libelous. Sign your letter. Be sure your heading includes your return address in case the editor wishes to check some matter with you.

Below, for study, are letters that have appeared in technical and professional journals. Notice that editors frequently provide a headline as an eye-catching capsule. The salutation varies with the publication. In the interest of conserving space, some periodicals omit the salutation and all publications omit the complimentary close. Do include both in the letter you send. In other respects, follow the style of the periodical to which you forward your letter.

Symmetrel not a Vaccine[2]

Dear Sir:

In C&EN for June 6 (page 25) "Symmetrel" amantadine hydrochloride is referred to incorrectly as an "oral flu vaccine."

Amantadine hydrochloride is not a vaccine but is a novel, synthesized drug taken orally, which acts by interfering with virus penetration of the host cells. It is the product of a long and intriguing study of adamantane chemistry and an extensive research program in antiviral chemotherapy.

In contrast, a vaccine is defined by Webster as: "A preparation of killed microorganisms, living attenuated microorganisms, or living fully virulent organisms that is administered to produce or artificially increase immunity to a particular disease."

Because vaccines are the traditional approach to antiviral therapy, and chemotherapy is a new concept, we recognize that the distinction may not be clear to everyone. However, since your readers have such high regard for your technical accuracy, especially on chemical subjects, I hope you will not think us presumptuous in calling the difference to your attention.

George H. Soule

Wilmington, Del.

Support the Theoretical Thinkers[3]

Regarding Greenberg's article (News and Comment, 24 June, p. 1724) on "Basic Research: the political tides are shifting" ... the old category of "basic" ("fundamental," "pure") research is not good because it means different things to different persons. Really there are two kinds of so-called basic science that must be considered—inscribed (trivial, limited)—and theoretical. Inscribed science is simple fact-finding science without any direct thoughts about established theories or without any new set of postulates in mind that might ultimately become a theory. When I try to learn the nutrient conditions that cause the tips of fungus filaments to lyse, I am engaged in inscribed (trivial) science because my thinking has not related the facts to established theory or to a new set of postulates that might develop into a theory. This kind of basic (inscribed, trivial) science is often on the same intellectual level as applied science. It is fascinating work because it satisfies one kind of curiosity. But this kind of non-applied research should not be confused with truly theoretical work like

---

[2] *Chemical and Engineering News,* July 18, 1966.
[3] *Science,* Oct. 7, 1966, No. 3, Vol. 154.

that which produced the 1 : 1 : 1 hypothesis, the operon hypothesis (both of which should now be called theories), or the like.

Of the different kinds of science, theoretical science is the kind that should receive unstinted financial support because theories give us command of knowledge whether we choose to use the knowledge in practical applications or in the advancement of science. A quick review of the history of physics, chemistry, or biology will support this contention.

Within theoretical science are those men whose genius and drive permit them to build the frameworks of new incipient theories, new postulational-deductive systems. But who, in politics or in science, has the wit to recognize these men while they are in the early stages of formulating their ideas? Probably only a few other scientists and these men are seldom in a position to grant financial support. The greatest problem in the support of science is not whether you support basic or applied science, but rather, how do you support truly theoretical work?

Ralph W. Lewis

Department of Natural Science
Michigan State University, Lansing

The following is an example of a letter that is critical of a previously published article. The author of the article that is criticized is then given an opportunity to reply.

Psychology Experiments Without Subjects' Consent[4]

On occasion a scientific paper may incidentally, and perhaps unintentionally, reveal information of more significance concerning practices and attitudes of scientists in a given field than it does about the subject under investigation. I refer to the article by Rokeach and Mezei, "Race and shared belief as factors in social choice" (14 Jan., p. 165).

On the basis of this report it appears that in psychology it is considered permissible to experiment upon job applicants without their permission or knowledge. The authors document this by their statement that "(the subjects) were under the impression that the procedures to which they were subjected were an integral part of a normal interview procedure, and they were totally unaware that they were participating in an experiment. . . ."

I protest that this represents an invasion of fundamental human rights, namely the right to privacy and the right not to be subjected to manipulation and experimentation without one's knowledge and consent. These rights are now well recognized in research, and some recent breaches by medical researchers have been heavily censured. . . .

It appears to me that one of the most fundamental aspects of a civilized culture is that the citizen may correctly assume that in ordinary day-to-day activity he will be treated with candor and dignity and that, in general, he can

---
[4] *Science,* April 1, 1966.

trust the individuals with whom he deals. The business world operates upon these assumptions; and to the degree that they are not observed, business is not civilized. In the practice of the learned professions there should be no place for activity which offends against these ideals. I hope that mine is not the only voice raised in protest against such practices in psychological research.

Samuel E. Miller

227 West Main Street
Abingdon, Virginia

The rebuttal follows:

It is true that behavioral scientists often engage in research with human subjects without first obtaining their informed consent. But I do not agree with Miller's contention that we invaded our subjects' fundamental human rights in the research we recently reported in <u>Science</u>. The moral issue is considerably more complicated than Miller makes it out to be. What is typically involved in making a decision about moral values, whether in or out of science, is not a choice between good and evil but a choice between two or more positive values, or a choice between greater and lesser evils. A person may, for example, have to choose between behaving honestly and behaving patriotically and behaving truthfully....

Much of the research on behavior would be scientifically worthless if the subject were to be first informed of its purpose.... The behavioral scientist faces a moral dilemma arising from the conflict between the high value he places on advancing scientific knowledge for the betterment of human welfare and the high value he places on his subjects' individual rights. I know of no simple moral principle which will resolve in advance this oft-encountered conflict. It is a dilemma which has to be struggled with anew every time it arises. Technical considerations aside, the particular research design the behavioral scientist settles on is the end result of weighing all the complex moral considerations.... There are certain lines of research I would like to pursue but which I have not pursued because I felt that the damage to experimental subjects would be too great and thus that the scientific knowledge would be gained at too great a cost.

How does the conscientious behavioral scientist resolve the value conflicts he continually faces? First and foremost, he must necessarily rely on the dictates of his own conscience to avoid experimental procedures which would result in the subject's humiliation or embarrassment, or physical or psychic pain. Second, he must at the same time distrust his own conscience as an infallible guide and check his moral judgments against those of his colleagues and friends. Third, he must adhere to the moral standards of his profession. The American Psychological Association, in 1953, published a booklet, "Ethical Standards of Psychologists," which represents its official policy with respect to psychologists' behavior in research and in other areas of professional conduct. Violators of this code are subject to censure or expulsion from the association.

If Miller's assumption, that prior consent is always the over-riding ethical requirement, were to be followed to its logical conclusion, many kinds of important research with human subjects would not be possible. Such a serious consequence should be evaluated within the context of a broader conception than that provided by Miller of the role behavioral science ought to play in advancing human welfare and individual freedom.

Milton Rokeach

Michigan State University
East Lansing

The Individualist in Science[5]

In a letter published in the January 1965 issue of PHYSICS TODAY (Page 134) I discussed what I felt to be a dangerous by-product of the nouveau riche status of the physics community—a by-product that had led to the present tendency toward team research and the accompanying establishment of "the path" toward scientific truth. I argued that these developments have been at the expense of risking the individuality and freedom of inquiry that have been so essential to scientific progress throughout the history of science.

The reaction to that letter (through private correspondence and conversation) has been divided into two disparate camps. One camp argues that the most efficient way to solve hard problems in science is to let the paths of inquiry be specified by a small number of leaders (who have previously shown themselves to be brilliant) and followed by the bulk of the scientific community. This camp argues that to let anyone strike off in any other direction to solve the important problems in physics would lead to chaos. The second camp, defending the individualist in research, argues that the history of science substantiates, in a great number of discoveries that have led to bona fide progress in science, that such advances were indeed due to the free thinking of individuals whose research did not conform to the established norms of their contemporaries.

If the scientific community is in full agreement with the contention of the former camp there is nothing more to say, since such a methodology is currently in practice. On the other hand, if the contention of the latter camp (which refutes the former through the experience of centuries of scientific research) is to be taken seriously, then the question remains: What can be done, within the structure of the present-day scientific community, to aid and encourage the individualist in science? Should the scientific community at least partially agree with the defense of the individualist in research, then the question is a most important one and should be given a great deal of serious thought.

As a first step in answering the question, it appears to me that we should examine one of the most important aspects of scientific research: the means of communicating its results. Since such communication is one of the chief roles of the American Physical Society and the other affiliates of the American Institute of Physics, I should like to make two proposals that I feel could induce a relaxation of the rigidity in present-day research and thereby aid the individualist.

My first proposal is based on the premise that the rigidity in present-day thinking could be relaxed if the journals would encourage publication of short

---
[5] *Physics Today,* **June 1966.**

communications relating to technical criticism and elaboration on research articles that have already been published in their own pages. I should like to propose that revival of the "Letters-to-the-Editor" column in The Physical Review and the introduction of such a column in the other AIP journals could be most conducive to progress in physics. In my opinion it is indeed unfortunate that a number of years ago the letters column was removed from The Physical Review. The subsequent establishment of Physical Review Letters—a most needed journal for the quick publication of short articles—did not replace the "Letters-to-the-Editor" column. Physical Review Letters is dedicated to new research, whereas the eliminated letters column was for criticism of and elaboration on articles that had appeared in The Physical Review.

My second proposal is perhaps more far-reaching, but just as important for obtaining maximal freedom for the individualist in research. I propose the establishment of one more AIP journal. Its aim would be somewhat different from the aims of other scientific journals, and it should not duplicate the kinds of articles that appear in the other AIP journals. The editorial policy of this journal should be to accept only articles that are in accord with some predetermined (and announced) criterion of completeness in the development of a definite "thesis" that is new and is exploited sufficiently in terms of mathematical development and comparison with experiment to make a bona fide, convicting case for the objective reader. Generally, then, such articles should be based on long, careful investigations, covering as many as possible of the avenues that follow from the central thesis. Because of the nature of such articles, the journal should not be inundated with countless manuscripts.

I hope the existence of such a journal would induce researchers to carry their investigations much further than is customary today before the results are written up. The journal could also relieve some of the burden on other journals that are now flooded with short articles on the initial stages of theoretical and experimental studies.

In addition to the requirement of completeness, the only criteria that should be used to judge the acceptability of a manuscript for the proposed journal should be originality, logical consistency and mathematical consistency. I propose that if the reviewing staff of the journal found an article that satisfied those four criteria—regardless of whether or not its central thesis departed from the thinking of the majority of researchers—the journal would be obliged to accept the article for publication.

I feel that such a journal would indeed aid the individualist by giving him a voice—on condition that he had not only speculated but also followed through with a rigorous mathematical and experimental development and had made sufficient comparison with the properties of nature. In addition to the possible relief that such a journal might provide for other AIP journals, it might also, in the long run, help reduce the tensions of the "publish-or-perish" policy in many institutions by relieving the pressure on scientists to publish results of their studies while the studies are still in the preliminary stages.

I hope that my fellow members of The American Physical Society, and members of the other AIP societies, will consider these proposals to aid the individualist by ensuring and facilitating, as far as possible, his freedom of inquiry in present-day science. I hope that this column might be used for further discussion of these and other suggestions to this end.

Mendel Sachs
Boston University

Scientific and technical people (contrary to the popular notion) are not humorless. Many like Albert Einstein and Harlow Shapley were and are capable of surges of great wit to illustrate or make a point. Technical professionals are no different than articulate spokesmen in other fields in their command of vivid irony to substantiate an argument or to clarify an issue by a perceptive play of wit. The editor's mail column in the professional and technical journals is frequently enlivened and enriched by such displays. The letter that follows is a good example.

Animal Testimony[6]

Greenberg's remarkable report (News and Comment, 16 Dec., p. 1424) on the new scientific challenge to the nation, "to teach an animal to speak in this decade," mentions that legal scholars, have been drawn into the project and are wrestling with the problem of the admissibility of animal testimony in legal proceedings.

Your readers may wish to know that legal scholarship has already solved that problem, as the attached opinion of the English Court of Appeal (Fictive) indicates.

Joseph D. Becker

40 Rector Street
New York, New York 10006

In the
Court of Appeal
REX v. BARKER
Welp, Cur and Bellow,
Justices

In this appeal, we are called upon to decide the extraordinary question whether a conviction for larceny must be set aside on the ground that the only evidence against the convicted appellant was the testimony of a dog.

At the threshold, we are confronted by a curious argument, advanced rather gruffly by counsel for the appellant, Mr. Collie. It is urged that our Admissible Evidence Act enables any "person" to testify in a judicial proceeding; that the dog, known as Spot, who was allowed to testify against the appellant, Barker, was manifestly not a person; and, consequently, that the conviction of Barker rests on inadmissible evidence and must be quashed.

[6]*Science,* 10 March 1967, p. 1195.

The argument has a certain superficial appeal, but the law is quite capable of dealing with sophistries. In other instances, our decisions have held that corporations, partnerships, public bodies, women, ships, and tuna fish are "persons" within the meaning of pertinent statutes. On the Continent, Professor Schnauze has collected the cases in his monumental work, Hunde und Recht, especially with regard to German shepherds, and concludes that dogs are persons, or at least "quasi-persons in a Wagnerian sense" (p. 627, translation). In America, a court has plainly held that a dog is man's best friend. People v. Mutt, 100 Tex. 1(1880) (Poodle, C.J.). We accordingly hold that Spot was a "person" within the meaning of our Admissible Evidence Act and was competent to testify.

The appellant next contends that Spot did not properly take the oath required of witnesses. The record does disclose that when Spot was asked by the bailiff whether he did "solemnly swear to tell the whole truth, etc." his answer was more of a growl than a clear affirmation. But Johnson's reaction to a dog walking on its hind legs is apposite: it is not done well; but you are surprised to find it done at all. A reasonable dog cannot be held to the same standards as a reasonable man. The oath was satisfactorily taken.

The appellant's final contentions are directed to the examination of Spot by the Counsel for the Crown, Mr. Terrier, that elicited critical testimony against Barker. The record discloses the following colloquy:

TERRIER: Now, Spot, do you see the thief anywhere in this courtroom?
SPOT: Grrr Grr.
TERRIER: Can you point him out?
SPOT: Grrr. Grr. (raising paw).
TERRIER: Let the record show that Spot pointed his right front paw in the direction of the accused.

The appellant moved to strike Spot's testimony as unresponsive because, it is asserted, Spot was not here answering questions but was merely scratching at fleas. The trial judge denied the motion. Judge Nimrod, with his broad experience in these matters, was plainly in a better position than we to determine whether Spot was answering or scratching. The judgment of conviction is AFFIRMED.

THE MEMORANDUM

The memorandum is one of the professional's most frequently used instruments of communication. He uses it to record information worth keeping; to circulate information he wants to share; to make a record of policies, decisions, and agreements; to inform members of his organization of new policies, situations, and activities; to report on a trip; and to summarize and record the business meetings of his committees.

It is appropriate to review the latter part of Chapter 4 on the format of memoranda. In mechanics, the memo is similar to the letter, but its tone is more impersonal. A major objective of the medium is to save time for both the reader and writer. Amenities are relinquished for conciseness; format and language are directed to move the message along. Memos are frequently written under the pressure of time, and the writer is required to analyze a message situation quickly and to formulate it succinctly. His analysis must reduce the subject (no matter how complex) to its substance, and this to a terse, single statement in his Subject Line. The elements that formed the synthesis must be organized into related components that lead the reader to the central idea or major purpose of the memo. Superfluous detail must be pruned from the ambient body of information to avoid confusing the reader. The reader is given direct, concise information and facts, with conclusions and recommendations (as appropriate) to provide clear but ample background for him to arrive at proper decisions and necessary actions. Consider the following attempt at a memorandum as an illustration of a typical situation calling for this medium.

LARRABEE PNEUMATIC INSTRUMENTS
Interoffice Memorandum

TO: Mr. S. Cedric Larrabee, President
FROM: Virgil S. Perry, Administrative Assistant
SUBJECT: Trip Report on the American Management Association Conference on Employee Personnel Problems

Due to very bad weather my plane was delayed leaving our airport and it arrived more than two hours and thirty three minutes late to New York. Instead of landing as scheduled at Kennedy International Airport, it was rerouted to LaGuardia. Fog and other foul-ups delayed us at least another hour and a half.

This brought me to the Waldorf Astoria at 10 PM, four hours beyond the six o'clock deadline for my hotel reservation. Just as I feared, all the rooms were gone and I had to settle for a second rate hotel six blocks from the Waldorf where the conference was being held. It rained the three days of the conference, adding to the fun of getting to and from the meetings.

Despite these frustrations and inconveniences the three days were not a washout, due mostly to a session on the third day conducted by Dr. S. I. Nitze, a consultant in industrial psychology. But I am getting ahead of myself. I shall detail all of the sessions even though all of them except for Dr. Nitze's were pretty much old hat.

The first session on Monday—(thank God for the coffee served through the excellent arrangements of the American Management Association—It knows how to run meetings!)—was on "Things to Watch for in Interviewing Salaried Staff," etc., etc., etc.

Young Mr. Perry is reliving the trying three days of his trip to New York to attend a conference. Old Larrabee, though a kind and sympathetic boss, cares little about the inconveniences his assistant suffered while broadening his background at a conference and cares less to have them recorded formally. All he wants to know is: What did you learn that we might want to consider applying at Larrabee Pneumatic Instruments? He wants that information short and sweet. Of course, he will not hesitate to let Virgil S. Perry know this. A little red in the face Perry tries again. He is a bright young fellow, quick to learn. His next effort old Larrabee wants shared with his division managers. That memo might be as follows:

LARRABEE PNEUMATIC INSTRUMENTS
Interoffice Memorandum

TO: Division Managers
FROM: Virgil S. Perry, Administrative Assistant to the President
SUBJECT: What Makes Good Employee Morale

I have just returned from an American Management Conference in New York. A session I found extremely valuable was on Improving Employee Morale, conducted by Dr. S. I. Nitze, a well-known industrial psychologist. Mr. Larrabee has suggested that I pass on to you a summary of Dr. Nitze's research and recommendations for improving employee morale.

Factors Important to Morale

Dr. Nitze stated that research has highlighted 14 factors as important in building morale:

1. Security
2. Interest
3. Opportunity for advancement
4. Appreciation
5. Company and management
6. Intrinsic aspects of job assignment
7. Wages
8. Supervision
9. Social aspects of job
10. Working conditions
11. Communication
12. Hours
13. Ease
14. Benefits

Please note that wages are halfway down the list. We should not jump to the conclusion that money is unimportant to an employee. Once his basic needs are satisfied through adequate pay, Dr. Nitze says, other non-monetary factors of an employee's job take on an ever increasing significance.

Recommendations for Improving Employee Morale

Dr. Nitze said that there are no simple ground rules. Every situation is unique—no two employees nor two companies are identical. However, psychological research on morale and employee attitudes have indicated the following recommendations:

1. Tell and show your employees that you are interested in them and would be glad to have their ideas on how conditions might be improved.

2. Treat your employees as individuals; never deal with them as impersonal variables in a working unit.
3. Improve your general understanding of human behavior.
4. Accept the fact that others may not see things as you do.
5. Respect differences of opinion.
6. Insofar as possible, give explanations for management actions.
7. Provide information and guidance on matters affecting employees' security.
8. Make reasonable efforts to keep jobs interesting.
9. Encourage promotion from within.
10. Express appreciation publicly for jobs well done.
11. Offer criticism privately in the form of constructive suggestions for improvement.
12. Train supervisors to think about the people involved insofar as practicable, rather than just work.
13. Keep your employees up-to-date on all business matters affecting them, and quell rumors with correct information.
14. Be fair.

Below is an example of what might be considered a routine memorandum whose purpose is to circulate information.

MEMORANDUM

February 18, 1969

TO: Laboratory Personnel

FROM: L. E. Morrison, Director of Engineering

SUBJECT: New Scales for Southwestern GM-100A McLeod Vacuum Gauge

We calibrated a Southwestern GM-100A McLeod gauge for use as a Laboratory Standard and sent the results to Southwestern Scientific Instrument Company for comments on the errors observed. Southwestern then advised us that its quality control personnel had recently discovered an error of about 2% in the 0-10 torr range on the scales for the GM-100A gauge and would furnish new scales for all GM-100A gauges in our possession.

The 2% error, of course, is not serious for most applications. However, laboratory personnel having a GM-100A gauge should obtain the new scale by contacting Mr. Tom Hatch at the Laboratory Supply Office, extension 4679 or 4680. The scales should be changed by no later than April 1. The old incorrect scales can be identified by the part number 68170 which is printed on the lower left corner of the scale plate.

The Memorandum For File

The Memorandum for File, which is ostensibly for the "file" of the writer, is a useful means for storage of information, conversations, events, observations, ideas, thoughts, hypotheses, or anything else worth (or thought to be worth)

preserving as a matter of record. The original or copies are often routed to persons who are concerned, or should be concerned, with the matter involved. At a future time when events have matured, the record on file is retrieved for action, further deliberation, or for a more direct and appropriate usage. Below is an example of this kind of memorandum.

TO : File July 22, 1968
FROM : H. J. Whitlow and S. A. Ross
SUBJECT : Application of Data on Surface Tension of Liquids to Problems of:
(a) Lubrication, (b) Cleaning and Salvage of Damaged Electronic Equipment.

The surfaces of liquids act as if they are in tension and become as small as is feasible with due allowance for the force of gravity and the effects of any container involved. A common manifestation is the raindrop. The measure of this tendency is called surface tension.

A drop of liquid touching a solid surface will either spread into a film or retain its identity as a drop. Drivers can notice these tendencies on windshields which are clean or dirty, and hence slightly greasy. Systematic investigation of spreading has shown that for a given surface all liquids with a surface tension below a given value—the critical surface tension—will spread; those above will not.

Compilation of data on critical surface tensions by the Naval Research Laboratory has led to several fruitful applications. Among these are procedures for keeping lubricants from moving around in delicate instruments such as watches, gyrocompasses where lubricant is needed in one place and undesirable in another, and procedures for displacing substances in the salvage of equipment. The latter saved the U. S. Navy on the order of millions of dollars after the fire aboard the USS Constellation, by providing general know-how especially on delicate electronic equipment.

Similar application of data on surface tension of liquids could prove to be appropriate to our procedures and products. Closer study of these data should be made.

The following reports are available, giving details of the above:

1. NRL Report 5680—Surface Chemical Methods of Displacing Water and/or Oils and Salvaging Flooded Equipment. Part 2—Field Experience in Recovering Equipment Damaged by Fire Aboard <u>USS Constellation</u> and Equipment Subjected to Salt-Spray Acceptance Test, H. R. Baker, P. B. Leach and C. R. Singleterry, Sept. 61.

2. The Control of Liquid Spreading by C. R. Singleterry, NRL Progress Report, July 64.

MEETINGS

Technical professionals take part in many types of meetings: in meetings of international and national conventions and congresses of professional and industrial societies; in meetings of civic, legislative, and social organizations; in meetings of committees and subcommittees of organizations and societies; in meetings of *ad hoc* groups; in meetings of formal and informal boards; in meetings of chapters, councils, and commissions; and in staff meetings and conferences of all types.

The professional's participation in meetings may require him to be responsible for several types of written material: (1) Notice of Meeting; (2) Agenda; and (3) Minutes of Meetings.

Notices

The announcement of a meeting should be sent soon enough to give the members of the group sufficient time to plan to attend or to designate an alternate. The announcement may be written as a memorandum or as a notice. In either format, the announcement includes all (or most) of the following items.

1. Identification of the group or subgroup sponsoring or holding the meeting.
2. Number of the meeting (if it is one of a series). For instance:

> Notice of the Third Meeting of the Subcommittee on Microwave Behavior of Ferromagnetics and Plasmas of the Communication and Television Group of the Institute of Electrical and Electronic Engineers

3. Date, time, and place of the meeting.
4. Name of the Chairman, speaker(s), special guests, and so forth, as appropriate.
5. Agenda and background material and papers.

The agenda is often separate from the announcement of the meeting, and is attached to the announcement as an enclosure. The agenda may not be fully organized at the time the notice is circulated. Instead of the agenda, the announcement may contain a brief paragraph stating the purpose of the meeting. Sometimes topics to be considered at the meeting will have background papers distributed and attached to the announcement. Below is an example of a notice of a meeting. The announcement could have taken the form of a memorandum by having the notice line placed at the Subject Line.

AMERICAN CHEMICAL SOCIETY
Division of Carbohydrate Chemistry

Notice of the fourth meeting of the working committee on Carbohydrate Metabolism

March 14, 1968 9:30 A.M.

Room 310 American Chemical Society Building

Subject for discussion—Glucose and Glycogen Metabolism

The full agenda will be mailed by March 1, 1968

S. O. Durakian, Chairman

Enclosures:
   Papers by Norman E. West
           Jacob Silverstein
           F. W. Pearson

Agenda for Meetings

Literally, agenda (plural: agendas) means memoranda of things to be done. The term has come to apply to the document that lists things to be done at a meeting. If the topics to be discussed are known before the announcement of the meeting is sent, the agenda may be incorporated into the notice or may serve as an enclosure. If the topics for discussion are not known when the notice is released, the agenda either may be sent later or may be distributed at the meeting. The meeting follows the sequence of the topics or tasks listed. If the agenda is prepared separately from the notice, it should have a heading that identifies the meeting. The agenda may also have the format of a memorandum. The following items are included in the agenda, as appropriate.

1. Identification of the group or subgroup sponsoring or holding the meeting.
2. Date of the meeting.
3. The word Agenda centered on the page, unless the agenda is constructed in the format of a memorandum, in which case it becomes the first word of the Subject Line. For example:

    Subject: Agenda for the Subcommittee on Intermediate Degree Programs at non-Doctorate-Granting Institutions.

4. List of topics, subtopics, and tasks (if any), numbered and lettered for ease of reference.
5. List of background materials or papers (if any).

Below is an example of an agenda.

AMERICAN INSTITUTE FOR POLLUTION CONTROL
Industrial Liaison Committee Meeting
United Engineering Center, New York, N. Y.
June 2, 1968
AGENDA

9:00	Introductory Remarks (P. J. Autrey)
9:05	Report on Education of Management (M. S. McAloo)
9:30	Report on Education of Operational Personnel (T. J. Wassink)
10:00	Discussion
10:30	Report on Pulp and Paper Industry (N. V. Resnick)
11:00	Report on Oil Refining (B. H. Glater)
11:30	Discussion
12:30	Lunch
2:00	Report on Food Processing in Canning Plant Operations (W. E. Bagnell)
2:30	Report on Steel Industry Water Pollution Control Operations (R. C. Attan)
3:00	Report on Plant Operation for Pollution Control of Inorganic Chemicals (E. C. Liss)
3:30	Report on Organic Chemicals and Petrochemicals (M. A. LaRue)
4:00	Review and Questions (P. J. Autrey)
4:30	Recommendations
5:00	Adjournment

MINUTES OF MEETINGS

As a professional, you will be called upon at intervals (more frequent than you wish) to serve as secretary of a committee. A chore that inexperienced secretaries find painful is keeping a record of what happened at a meeting. This chore can become less burdensome if you are aware of what and how to record.

The minutes are not a verbatim transcript of the discussions, but are a concise, clear summary of the major topics discussed and of the conclusions and recommendations reached, as well as what the committee as a whole or what individual members agreed to do or were assigned to do. Lengthy minutes take much time to prepare, too much time to check, frequently cause resentment in persons who think they have been misquoted, and (in the end) are not read by most members of the group. Brief, well-written and well-organized minutes provide the formal record of the group's meeting and its results. Before minutes are formally distributed they should be checked for accuracy, completeness, and consensus by the chairman and participants.

The recording of minutes should not be assigned to the youngest, least experienced member of the group. The inexperienced secretary may inadvertently include statements or matters—even though dealing with crucial issues of the meeting—best left unrecorded. A consensus is often reached after heated debate in which strong feelings and emotions are involved. To place on record statements that manifestly might be true but that are also provokingly virulent in their tactlessness is to invite bitter discord, wrangling, recriminations and counter-recriminations. Consider, for example, the following entry in a Minutes:

> The Executive Committee examined the recommendations of the Society's Goals Committee. After a close vote, the Executive Committee rejected the Goals Committee recommendations and returned its report for further study. The Goals Committee was instructed to be more concerned with immediate problems and to concentrate less on long range goals. The Goals Committee is to come back with a new report and recommendations no later than three months from receipt of these instructions.

A more tactful and equally valid entry for the situation might have been phrased as follows:

> The Goals Committee Report was discussed. Because several critical problems have materialized since the Goals Committee received its assignment, it was decided to request the Goals Committee to reshape its focus to include immediate goals along with the previously assigned long range goals. The Goals Committee was given an additional three months to provide the augmented approach to its study.

The minutes of a meeting or conference should contain the following elements, but not necessarily in the order listed:

1. The name or identification of the committee or group and sponsor, if appropriate.

2. The date(s), of the time and place of the meeting.

3. A list of members and officers attending; a list of persons attending by invitation for special discussions; and a list of members who did not attend. (The list of members who did not attend varies with practice.) In the minutes for a group with a stable membership, the record of attendance may be shown by listing only the names of those absent. For example:

> All members of the Executive Committee were present except Dr. James V. Foote and Professor Eric Van Olphen.

4. The time when the meeting was called to order, recessed (if applicable), and adjourned is noted.

5. The text of the minutes is arranged by topics. Paragraphs may be identified by number and title to correspond with the agenda items. This type of grouping aids quick and easy reference. In minutes longer than a page, it is

useful to type the participants' names in capital letters whenever they appear in the text.

6. Interim executive actions between the previous meeting and the present meeting are recorded.

7. Discussions are summarized (in sentence form) in clear concise statements. Responsibility for action, as called for, is clearly indicated.

8. The signature of the secretary or the initials of the person preparing the minutes is placed at the end.

9. Additional information may be included—items such as the date and place of the next meeting, assignment of specific duties, or items that were added to the previously circulated agenda.

Here are two examples of minutes of meetings for study: One covers the meeting of a subcommittee, and the other covers the meeting of the board of directors of a giant professional society.

MINUTES OF
SHIELDING CROSS SECTION SUBCOMMITTEE
of the Physics Problems of Reactor Shielding Committee
of the National Nuclear Energy Institute

Saturday, April 19, 1968   Conference Room, Administration Building
Oak Ridge National Laboratory, Tennessee

Subcommittee Members Present were: Absent were:

Jasper A. Cairns, ORNL, Chairman   C. K. Knox, BN
Maurice Magoun, Los Alamos Scientific Laboratory   L. R. Tennis, KAPL, GE
P. V. Schmidt, Westinghouse Electric Corporation
Byron Spiegel, General Atomic
Blaine K. Steere, Argonne National Laboratory
Lise J. Weisman, MIT, Secretary

1. CALL TO ORDER
   The Subcommittee was called to order at 9:00 A.M.
2. MINUTES OF DECEMBER 5, 1967 MEETING
   Approved as written.
3. SUBCOMMITTEE PERSPECTIVE
   To provide a proper perspective for the mission of this subcommittee, Dr. Cairns presented a review of the state of the art and practice of shielding cross sections. Progress of the work of this Subcommittee was established one year ago and was also reviewed for the two new members.
4. ADVISORY ACTIVITIES
   This Subcommittee has been formally or informally invited to provide advice and counsel for the following:
        a. The Department of Defense shielding activities
        b. National Bureau of Standards
        c. The Aldermaston, England establishment

    d. The European Neutron Center at Saclay, France
    e. The Japanese Nuclear Data Committee

The membership of this Subcommittee voted favorable consideration of these requests and went on record to urge the National Nuclear Energy Institute to encourage broadly this type of cooperation on the part of the Institute.

5. REACTIONS AND DATA

A list of data, interactions, and functions of special interest to shielding has been prepared by a special working group under cognizance of the Subcommittee. The listing emphasizes neutron induced processes. The Subcommittee voted to disseminate the listing. (The list is attached to the Minutes as Exhibit A.)

6. GAMMA RAYS AND OTHER PARTICLES

After discussion, it was agreed that gamma ray induced interactions must be of interest to shielding, but that work on the gamma ray problem should be deferred until the next meeting when the chairman of the working group (C. K. Knox) would be present. It was also noted that the (p, n) reaction can be important to shielding water reactors. It was further decided that following the ongoing study recommendations will be made regarding (x, n) reactions.

7. ESTABLISHMENT OF INITIAL DATA PRIORITIES

After discussion, the following mechanism was decided upon for establishing neutron data priorities: The Secretary of the Subcommittee (L. J. Weisman) will collect from the other members their initial recommendations. Also, responses from contractors to be polled (see attached list, Exhibit B) will be funneled to the Secretary. She will consolidate and roughly order the recommendations. Then the consolidated list will be circulated among the Subcommittee for comment on the order and on location of relevant data. The Secretary will then revise the list for submission of the Subcommittee at its next meeting.

8. WORK ASSIGNMENTS

J. A. Cairns—Data priorities
C. K. Knox—Planning gamma data requirements
B. Spiegel—Format requirements
B. K. Steere—Recommendations for storage and processing data
P. V. Schmidt—Liaison with DOD and NBS
M. Magoun—Liaison with foreign sources
L. R. Tennis—Contractor poles
L. J. Weisman—List of recommended data priorities

9. NEXT MEETING

September 27, 1968 at Argonne National Laboratory, B. K. Steere, Host

10. ADJOURNMENT

Meeting adjourned 4:45 P.M.

Lise J. Weisman
Secretary of the Subcommittee

2 Encls.

Exhibit A, List of data, interactions and Functions of Special Interest to Shielding.
Exhibit B, List of Contractors to be Polled.

MINUTES OF THE AMERICAN CHEMICAL SOCIETY BOARD MEETING

June 3, 1966

The Board of Directors of the American Chemical Society met at ACS headquarters, Washington, D. C. at 10 A.M. on June 3. Arthur C. Cope, Chairman, presided. The following Directors were present: A. M. Bueche, R. W. Cairns, Arthur C. Cope, Milton Harris, W. O. Milligan, John H. Nair, Charles G. Overberger, Byron Riegel, John C. Sheelhan, William J. Sparks, and Charles L. Thomas.

The following were present by invitation for part or all of the regular sessions: Dale B. Baker, R. H. Belknap, Gordon Bixler, Boris E. Cherney, Alden H. Emery, R. N. Hader, Arthur B. Hanson, E. G. Harris, Jr., R. E. Henze, R. K. K. Jones, Richard L. Kenyon, Joseph H. Kuney, R. V. Mellefont, Stephen T. Quigley, R. L. Silber, James H. Stack, B. R. Stanerson, Fred A. Tate, R. M. Warren, and Ralph F. Wolf.

1. The minutes of the meeting of March 25, 1966, were approved (see C&EN, May 23, Page 62).

AD INTERIM ACTIONS

2. VOTED that the following ad interim action of the Board of Directors be confirmed:

   VOTED to approve the draft of the minutes of the meeting of the Board of Directors on March 25, 1966 as transmitted on April 12.

3. VOTED that the following ad interim actions of the Committee on Awards and Recognitions be ratified:

   VOTED that the Committee on Awards and Recognitions, acting for the Board of Directors, accept the rules and regulations for administration of the biennial American Chemical Society Award in Fertilizer Chemistry as proposed by the sponsor, the National Plant Food Institute.

   VOTED that the Committee on Awards and Recognitions, acting for the Board of Directors, accept the rules and regulations for the administration of the American Chemical Society Award in the Chemistry of Plastics and Coatings as proposed by the sponsor, The Borden Company Foundation, Inc.

   VOTED that the Committee on Awards and Recognitions, acting for the Board of Directors, prepare a congratulatory scroll to be presented to the Czechoslovakian Chemical Society at its centennial celebration July 4-7, 1966 in Prague. (Dr. John C. Sheehan presented the scroll on behalf of the Board of Directors at the meeting.)

4. VOTED that the following ad interim action of the Committees on Finance and on Publication be ratified:

   VOTED that the Committees on Publications and on Finance, acting for the Board of Directors, authorize the conversion of CHEMICAL ABSTRACTS to a weekly publishing cycle in 1967, with approximately one half of the range of subject matter to be covered each week.

5. VOTED that the following ad interim action of the Committee on Grants and Fellowships be ratified:

   VOTED that the Committee on Grants and Fellowships, acting for the Board of Directors, on recommendation of the Petroleum Research Fund

Advisory Board, approve the allotment of funds for the PRF grants and awards listed in Tables I through V.

REPORTS

6. VOTED to receive the reports of the officers of the society.

7. VOTED to receive the reports of the following committees and Boards: Awards and Recognitions; Corporation Associates; Education and Students; Finance; Grants and Fellowships; Public, Professional, and Member Relations; Publications; Special Committee of the Board of Directors for the Chemical Abstracts Service; Chemistry and Public Affairs; Investments; To Study Plans for ACS Employees to Purchase Annuities; Patent Matters and Related Legislation (joint with Council), the Petroleum Research Fund Advisory Board and the Board of Trustees for Administering the Group Life Insurance Plan for ACS Members. Those reporting more than progress are printed on page 58-63.

8. VOTED to receive the reports of the Controller and the Chairman of the Committee on Investments.

AWARDS AND RECOGNITIONS

9. By secret ballot, VOTED that the Priestley Medal for 1967 be awarded to Ralph Connor.

10. On recommendation of the Committee on Awards and Recognitions, VOTED to change the name of the ACS award in Industrial and Engineering Chemistry sponsored by Esso Research and Engineering Company to the E. V. Murphree Award in Industrial and Engineering Chemistry sponsored by Esso Research and Engineering Company.

11. On recommendation of the Committee on Awards and Recognitions, VOTED to authorize extension of the biennial Roger Adams Award in Organic Chemistry for five award years (1969-77) and to thank the donors (Organic Syntheses, Inc., Organic Reactions, Inc., and the Division of Organic Chemistry) for increasing the honorarium from $5000 to $10,000 beginning in 1967.

CONSTITUTION AND BY-LAWS

12. VOTED to confirm the amendment of By-Law III, Sec. 3(a)(2) providing for the inclusion of the immediate Past President on the Council Policy Committee (see Council Minute 9, C&EN, May 23, page 74).

FINANCE

13. On recommendation of the Committee on Finance, VOTED to authorize the Controller to initiate the necessary actions to discontinue the Revolving Fund concept which was set up originally to take care of expenditures that were not of an annual type and to budget those activities involved as current revenue and expense functions.

FINANCE-PUBLICATIONS

14. On recommendation of the Committees on Finance and on Publications, VOTED that a new publication in the field of environmental chemistry be established to include papers on original research and engineering development in fields of chemistry concerned with man's environment and under appropriate

circumstances this journal will include critical reviews and surveys on selected topics of current interest to scientists and engineers working in various segments of the field of environmental chemistry, reports of scientific and technical meetings written by experts in the field concerned, and a modest amount of staff-written interpretive treatment of developments in research, industrial activity, and legislative and political activity at the federal, state, and local level where such activity has a professional connotation to workers in the field.

15. On recommendation of the Committee on Finance and on Publications, VOTED that the 1967 prices for the alerting service to the Chemical Abstracts Service (Polymer Science and Technology (POST)) be set as follows and that magnetic tape copies be made available to subscribers at $500 per year for each section, with the subscriber furnishing the tape:

	POST-J*	POST-P**	TOGETHER
Base price up to 25 scientists with an additional $50 for each additional group of 25 or less on the subscriber's staff to a maximum of:	$1200	$1000	$2100
	1700	1500	3100

PUBLICATIONS

16. On recommendation of the Committee on Publications, VOTED to appoint Harold Hart to succeed Ralph Shriner as editor of CHEMICAL REVIEWS, effective Jan. 1, 1967.

CORPORATION ASSOCIATES

17. On recommendation of the Committee on Corporation Associates, VOTED to redefine the committee's purposes in broader perspective as indicated in its report (see page 59) and to change its status from that of a Standing Committee to a Special Committee to permit the inclusion of persons who are not members of the Board of Directors.

EDUCATIONAL AND SCIENTIFIC EXPOSITION

18. On recommendation of the Committee on Finance, VOTED that the ACS sponsor and manage, on terms to be agreed on between the Society and the California Section, an educational and scientific exposition at the 155th National Meeting in San Francisco, March 31 to April 5, 1968.

19. On recommendation of the Committee on Finance, VOTED that a general policy decision on ACS sponsorship and management of educational and scientific expositions be deferred until after the results of the Miami Beach exposition (April 9-14, 1967) have been analyzed.

MEETINGS

20. In collaboration with the Council Policy Committee, VOTED to approve the following schedule of business sessions at the 152nd national meeting this fall:
    Board of Directors, Sunday, Sept. 11
    Council Policy Committee, Monday afternoon, Sept. 12
    Council, Tuesday morning, Sept. 13

Standing committees, executive sessions—to be set by committees but at least one session completed prior to the start of the CPC meeting

Standing committees, open sessions—to be set by committees

21. On recommendation of the Council, VOTED that the 1967 winter meeting scheduled for Kansas City, Mo., Jan. 15-20, be cancelled.

22. On recommendation of the Council, VOTED that the date of the 1969 spring meeting in Minneapolis be changed from March 23-28 to April 13-18.

23. On recommendation of the Council, VOTED that the 1972 spring meeting be held in Los Angeles, Calif., April 9-14.

MISCELLANEOUS

24. On recommendation of the Committee on Chemistry and Public Affairs and on Finance, VOTED to grant up to $10,000 for a conference on environmental improvement, including publication costs for such reports as may result from the conference.

25. VOTED to confirm the appointment of an ACS representative to the U. S. National Committee of the International Association on Water Pollution Research (Council Minute 28, March 27, 1966) and to approve payment of the required membership dues.

26. On recommendation of the Committee on Chemistry and Public Affairs and with the endorsement of the Committee on Public, Professional, and Member Relations, VOTED that the American Chemical Society endorse the adoption of the metric system in the United States.

The meeting adjourned at 5 P.M.

B. R. Stanerson, Secretary

PART III

APPENDIX

INDEX TO GRAMMAR PUNCTUATION, AND USAGE

This section is not meant to be a text on grammar and usage. Its purpose is to provide a ready and convenient annotated index of some of the more common problems the technical correspondent will meet in his letter writing requirements. Fuller treatment of some of these problems will be found in the listing of the Selected Bibliography that follows this Index.

A, An. *A* is used before words beginning with a consonant (except *h* when it is silent), before words beginning with *eu* and *u* pronounced *yu,* and before *o* pronounced as in *one. An* is used before all words beginning with a vowel sound and words beginning with a silent *h*.

a	*a*	*a*	*an*	*an*
gradient	halogen	eugenol	axis	hour
lobule	helix	onetrack	ohm	honor
solute	humus	univalve	M	heir
triode	hydrate	uropod	input	herb

Abbreviations. The technical person has a penchant for abbreviations. Abbreviations are appropriate in handbooks, compilations, and tables—works in which the saving of space is important. Abbreviations in correspondence should be used sparingly and only if they will convey meaning. Practice today is to write out the days and months both in the date line and in the body of the letter. Similarly, cities, states, and countries are written in full. A few prominent exceptions are Washington, D. C., St. Louis, and U.S.A. There are some standard abbreviations in correspondence practice—Dr., Mr., Mrs., and Messrs. However, Professor, Senator, President, and Reverend are not abbreviated. Dictionaries give frequently used abbreviations in their main listing of words; some also provide a special list of abbreviations in the back section. Periods are used with abbreviations with the exception of acronyms: for instance, NBS, NSF, AEC, UNESCO, and IEEE.

Above. Grammatically, this word is most frequently an adverb (the purchase order identified *above* [above modifies the verb *identified*]) or a preposition: *above* the ionosphere. *Above* is sometimes used as an adjective (the *above* purchase order) or as a noun: the *above* is requested. Many competent writers avoid using *above* except as a preposition because the sentence has a stilted construction and the thought transmitted tends to be ambiguous, since the reference of *above* (especially as a noun) is vague. *Above* as an adjective or adverb, in the context previously indicated, should be used with discretion and specificity. I would avoid its use as a noun entirely.

Accept, Except. *Accept* (verb) means to take when offered, to receive with favor, to agree to. *Except* (verb) means to exclude or omit; *except* (preposition) means with the exception of, but. He was *accepted* into Sigma Xi, he was the only one of four candidates *excepted* from Sigma Xi. All *except* him were admitted.

Access, Excess. *Access* means approach, admittance, admission, as: to gain *access,* easy *access. Excess* means that which exceeds what is usual, proper, just, or specified; as *excess* of grief.

Adapt, Adopt. These two words are frequently confused, although their meanings are entirely different. *Adopt* means to take by choice into some sort of relationship as: he *adopted* a child. *Adapt* means to modify, to adjust, or to change for a special purpose, as: he *adapted* the cam to provide a slower movement.

Adjective. An *adjective* is the part of speech that modifies or limits nouns and pronouns. Its purpose is to clarify for the reader the meaning of the word it modifies. The adjectives in the expressions *radioactive* substance, *alpha* particle, *heat-sensitive* material, *wire-wound* resistor, and *diastereoisomeric* ester provide more meaning to the reader than the nouns by themselves. Adjectives are classified as demonstrative, descriptive, limiting, and proper. A demonstrative adjective points out the kind of word it modifies: *this* instrument; *these* emissions; *that* solvent; and *those* techniques. A descriptive adjective denotes a quality or condition of the word it modifies: an *automatic* acid analyzer; a *polargraphic* accessory; a *broken* test tube; and a *higher-flow* rate. A limiting adjective designates the number or amount of the word it modifies: *15%* nickel nitrate solution; a *150* millimeter beaker; *400* grams of citric acid; and a *10* millimeter-per-minute flow. The articles *a, an,* and *the* are limiting adjectives. A proper adjective is one derived originally from a proper name: *McLeod* gauge; *Californian* jade; *Brownian* movement; and *Johnson* noise. Proper adjectives that have lost their sense of origin are written in lower case: *india* ink; *macadamized* road; and *angstrom* unit.

Adverbs. An *adverb* is a word, phrase, or clause that modifies a verb, an adjective, another adverb, or an entire sentence. Most adverbs are adjectives plus the ending *ly :* fina*lly,* correc*tly,* comple*tely,* and easi*ly.* Some adverbs, derived from old English, have no special adverbial sign: *now, quite, since, then, there,* and *where.* Some adverbs have the same form as adjectives: bad, better, cheap, deep, early, even, fair, first, hard, last, loose, second, tight, well, and wrong. Most of these adverbs also have an *ly* form.

Adviser, Advisor. Both spellings are acceptable. What adds to the confusion is that *adviser* is used more frequently, but the spelling of *advisory* is the correct one.

Affect, Effect. These words are frequently confused because many persons pronounce them to sound alike. *Affect* is almost always a verb, meaning to influence or to make a show of.

Improper diet *affected* his health.
He *affected* a French accent.

Affect as a noun is used as a term in psychology pertaining to feeling, emotion, and desire as factors in determining thought and conduct. *Effect* is rarely used as a verb. As a noun, it means result or consequence.

The *effect* of heat on the chemical reduced its absorbance capacity.
The *effect* of the new design was a saving in time and money.

Effect as a verb means to bring about.

The new design *effected* a great saving.

Agree to, Agree with. These are idiomatic expressions. You *agree to* things, and *agree with* persons.

He *agreed to* the contract.
He *agreed with* the client.

Although, Though. Both words often are used interchangeably to connect an adverbial clause with the main clause of a sentence to provide a statement in apposition to the main statement, but one that does not contradict it. *Although* is preferable to introduce a clause that precedes the main clause; *though* for a clause that follows the main clause.

Although we were short of the test equipment, we managed to check out the system.
We managed to check out the system, *though* we did not have all the test equipment.

Among, Between. *Among* denotes a mingling of more than two objects or persons; *between,* derived from an old English word meaning "by two," denotes a mingling of two objects or persons only.

Among several designs, his was the most practical. The instrument's accuracy was *between* ±1 and ±3%.

In current usage, *between* is used where the meaning denotes more than two objects or persons:

disagreement *between* the bidders
to choose *between* courses

Amount, Number. *Amount* refers to things or substances considered in bulk; *number* refers to countable items as individual units:

the *amount* of acid in the vial (but the *number* of vials of acid)
the *amount* of wheat in the barn (but the *number* of bushels of wheat)

And/or. This expression is commonly used in agreements, contracts, and other legal or business writing to show there are three possibilities to be considered.

He offered his house *and/or* automobile as security for the loan. (He offered his house, or his automobile, or both as collateral.)

Apostrophe. The most common use of the *apostrophe* is to denote possession and ownership:

Dr. Willoughby's hatbox
the formula's deviation
the corporation's president
the book's index

The *apostrophe* is used to indicate omission of one or more letters in a contracted word or figure:

'70 for 1970
I've for I have
didn't for did not

The *apostrophe* is used in the plural form of numbers, letters and words:

The *1970's*
Programming is spelled with two *m's.*
The second of the two *which's* is not necessary in the sentence.

The *apostrophe* is not used with the possessive pronouns:

his, hers, ours, theirs, yours, its

Appendix. The plural most commonly used is *appendixes.* But the purists prefer *appendices.*

Articles. There are three articles in the English language: *a, an,* and *the. A* and *an* are indefinite articles; *the* is the definite article.

Assay, Essay. *Assay* means to test; *essay* means to attempt. *Assay* also has the meaning of analyzing or appraising critically.

Await, Wait. *Await* is always a transitive verb (that is, one that is used with an object to complete its meaning). *Wait* is often intransitive, but not always.

I *await* your reply.
The agency *awaited* the proposal (expected it).
I *wait.*
I *wait* your decision.

Awhile, A While. *Awhile* is an adverb; *a while* is a noun phrase; it is often used as the object of a preposition.

The gas flame burned *awhile.*
For *a while,* the flame burned.

Because. *Because* is used to introduce a subordinate phrase or clause providing the reason for the statement in the main clause.

Because the temperature of the gas became so high, copper piping could not be used.

When a sentence begins with *The reason is* or *The reason why . . . is,* the clause containing the reason should not begin with *because* but with the word *that.* *Because* is an adverb; *that* is a pronoun. The linking verb *is* requires a noun clause rather than an adverbial clause.

Beside, Besides. *Beside* means near, close by, by the side of. *Besides* means in addition to, moreover, also, aside from.

The technician placed the panel housing *beside* the instrument.
Besides tensile strength, the alloy had other desirable properties.

Brackets [ ]. *Brackets* are used whenever it is necessary to insert parenthetical material within other parenthetical material. *Brackets* are also used to make corrections or explanations within quoted material.

But. *But* is the coordinating conjunction used to connect two contrasting statements of equal grammatical rank. It is less formal than the conjunctions *however* or *yet,* and is more emphatic than *although.*

Not an atom *but* a molecule.
The movement was slow *but* smooth.
The crew worked industriously under his supervision, *but* the minute he turned his back it sloughed off.

Can, May. *Can* expresses the power (physical or mental) to act; *may* expresses permission or sanction to act. In informal and colloquial usage, *can* is frequently substituted for *may* in the sense of permission, but this substitution is frowned upon in formal writing. *Could* and *might* are the original past tenses of *can* and *may.* They are now used to convey a shade of doubt or a smaller degree of possibility.

We *can* meet your rigid specifications.
You *may* apply for sabbatical leave at the end of the seventh academic year.
The chronopotentiometric method *might* be more susceptible to interference.
The strike *might* cripple the industry.

The use of *could* suggests doubt or a qualified possibility.

It is remote that under such conditions that the contractor *could* meet his delivery schedule.

Capitalization. The rules for capitalization in correspondence are the same as in other formal writing. Use *capitals* for:

1. Titles, geographical places, and trade names:
   Origin of Species
   Ross Sea, Antarctica
   Freon, Polaroid, Teflon, Scotch tape

2. Proper names and adjectives derived from proper names, but not words used with them:
   Lower Devonian System, Pliocene Period
   Gaussian, Coulombic, Einstein's theory of relativity, Babinski's reflex, Boyle's law

   Words derived from proper nouns and which, through long periods of usage, have achieved an identity within themselves are no longer capitalized. For example:
   pasteurized
   petri dish
   bunsen burner
   ohm, ohmic drop

3. Scientific names of phyla, orders, classes, families, and genera:
   Decapoda
   Megaloptera
   Colymbiformes
   Urochorda

Catalog, Catalogue. If you prefer the second spelling, you are a traditionalist fighting a losing battle.

Cheap. Is both an adjective and adverb.

Circumlocution. A long word meaning wordiness.

Wordy:	On the basis of the foregoing discussion it is apparent that . . .
Revised:	This discussion shows . . .
Wordy:	in this day and age
Revised:	today
Wordy:	Dr. Zinowski is studying along the line of cholecystography.
Revised:	Dr. Zinowski is studying cholecystography.
Wordy:	at the present time
Revised:	now
Wordy:	in the normal course of procedure
Revised:	normally, usually, ordinarily
Wordy:	We hope you will answer in the affirmative.
Revised:	We hope you will say yes.

Censor, Censure. *Censor* means to delete or suppress; *censure* means to condemn or blame.

Center About, Center Around, Center On, Center Upon. In formal writing the idiom is *center on* or *center upon;* in less formal writing, *center around* is the idiom.

Collective Noun. A collective noun is one whose singular form carries the idea of more than one person, act, or object: army, class, corporation, crowd, dozen, flock, group, herd, majority, personnel, public, remainder, and team. When the collective noun refers to the group as a whole, the verb and pronoun used with it should be singular. When the individuals of the group are intended, the noun takes a plural verb or pronoun.

> The committee *is* here.
> The committee *were* unanimous in disapproval.
> The corporation *has given* proof of *its* intention.

The plural of a collective noun signifies different groups.

> Herds of deer and cattle *graze* in the upper valley.

Colon. The *colon* is a punctuation mark of anticipation to indicate that the material that follows the mark will supplement what preceded it. Use the colon:

1. After the salutation of a letter:

   Dear Mr. McGraw:
   Gentlemen:

2. In memoranda following the TO, FROM, and SUBJECT Lines:

   TO:
   FROM:
   SUBJECT:

3. To introduce enumerations, usually with *as follows, for example, the following,* etc.

   The components are *as follows:*

4. To introduce quotations:

   The statement read:
   We the undersigned members of the corporation believe:

5. To separate hours and minutes:

   The plane arrived at 10:33 A.M.

Comma. The modern tendency is to avoid using commas except where they are needed for clarity of meaning. Use commas:

1. *To set off nonrestrictive modifiers or clauses.* A word, phrase, or clause that follows the word it modifies and that restricts the meaning of that word is called a restrictive modifier or restrictive clause. Restrictive clauses are *not* set off by commas. But when a modifier merely adds a descriptive detail (gives further information), it is called *nonrestrictive* and is set off by commas. If the modifier is omitted, will the sentence still tell the truth or offer the meaning that you intend? If it does, then the clause or modifier is nonrestrictive and should be

set off by commas. If it does not tell the truth and if it does not give the meaning intended, you must not use commas.

Example:

> Our chief engineer, who received his Ph.D. degree from California Institute of Technology, is now in our New York office.

The clause, "who received his Ph.D. degree from California Institute of Technology" is not necessary to the sense intended. Therefore, the clause is set off by commas. However,

> Our chief engineer who checked the computations of the experiment does not agree with the conclusions reached.

The clause "who checked the computations of the experiment" is necessary to the sense of the sentence. It is *restrictive* and therefore should *not* be set off by commas.

2. With explanatory words and phrases or those used in apposition (as an adjunct term):

> To meet the deadline, Mr. Cole, our president, supervised the preparation.

3. With introductory and parenthetic words or phrases, such as *therefore, however, of course,* and *as we see it:*

> As you are aware, absorbance changes least with time at 705 Mμ. The action is, of course, necessary.

4. In inverted sentence construction where a word, phrase, or clause is out of its natural order:

> That the result was unexpected, it was soon apparent.

5. To avoid confusion by separating two words or figures that might otherwise be misread:

> In 1969, 9600 employees still worked at the old site.

6. To avoid confusion by separating words in a series:

> The client manufactures metal furniture, cutlery, trimmings, and steel garden implements.

7. To make the meaning clear when a verb has been omitted:

> I covered the door, John covered the hallway, Bill covered the windows, and Tom, the remaining exit.

8. To separate two independent long clauses joined by a coordinating conjunction (*and, but, for, yet, neither, therefore, or, so*):

> Analog recorders of 0.1% accuracy are available, but frequent maintenance is often required to hold this tolerance.

(A comma is not used when the coordinator clauses are short and are closely related in meaning: A panel of experts might be chosen

from particular areas of specialization and their report would be published.)
9. To separate a dependent clause or a long phrase from its independent clause:

> For complex mixtures of acids such as those found in physiological fluids, a five-chamber concave gradient is generated.
>
> Although good recoveries of both methyl pyruvate and methyl lactate could be obtained at column temperatures below 100°C., there was no indication that such recoveries could be achieved when either of the acids were in excess.

10. To separate the day of the month from the year:

> December 21, 1916.

Usage is divided when the day of the month is not given, but usual practice is to omit the comma:

> December 1916.

In constructions in which the day precedes the month no comma is used:

> 21 December 1916.

11. To separate town from state or country when they are written on the same line:

> Salt Lake City, Utah
> Washington, D. C.
> Stockholm, Sweden
> Bethesda, Montgomery County, Maryland

12. In figures to separate thousands, millions:

> 436,211
> 5,784,093

Compare, Contrast. There is some confusion in the application of these two words because *compare* is used in two senses: (1) to point out similarities (used with the preposition *to*); and (2) to examine two or more items or persons to find likenesses or differences (used with the preposition *with*). *Contrast* always points out differences.

Complement, Compliment. *Complement* (noun) means the number or amount that adds up to the whole; *complement* (verb) means to make complete, supplement, or supply a lack in. *Compliment* (noun) means praise or commendation; *compliment* (verb) means to praise or to show regard for.

Compound Predicate. A compound predicate consists of two or more verbs having the same subject. It is often used to avoid the awkward effect of repetition of the subject or the writing of another complete sentence.

> The Emperor Van de Graaff accelerator is precisely controllable. Another feature is that it is easily variable. It is also continuous.

A more economical and smoother way of saying the above sentences is to combine them with a *compound predicate*.

The Emperor Van de Graaff accelerator *is* precisely *controllable, easily variable,* and *continuous.*

Compound Subject. Two or more elements serving as the subject of one verb are called a *compound subject. Compound subjects* require the plural verb form.

The deposition time and amount of carbon available in the gases determine the resistance value in film resistors.

Continual, Continuously. *Continual* means repeated at frequent intervals; *continuously* means uninterrupted.

Council, Counsel, Consul. These three words are often confused. A *council* is an advisory group; *counsel* means advice and, in law, it means one who gives advice; a *consul* is an official representing his government in a foreign country.

Dangling Modifier. A dangling modifier is usually a verb form (often a participle) that is not supplied with a subject to modify. It "dangles" because it has no word to which it can logically attach. Infinitive and prepositional phrases may also be dangling.

Wrong: Calibrating the resistance oftthe thermistor through the temperature range of 17° to 19° Centigrade, a value of 4.0 ± 0.1°C.1 molar was obtained.

Revised: Calibrating the resistance of the thermistor through the temperature range of 17° to 19° Centigrade, we obtained a value of 4.0 ± 0.1°C.1 molar.

Wrong: After adjusting the valves, the engine developed more power.

Revised: The engine developed more power after the valves were adjusted.

Or: After adjusting the valves, the mechanic found the engine developed more power.

Wrong: To write a program for our computer, it helps to know Fortran.

Revised: To write a program for our computer you would find it helpful to know Fortran.

Better: A knowledge of Fortran would be helpful in writing a program for our computer.

Wrong: Near Kamchatka, Alaska, Figures 11 through 17 show the typical appearances in both the television pictures and the infrared data, of several stages in the life cycle of the cloud vortices of a cyclonic storm.

Revised: Figures 11 through 17 show typical appearances in both television pictures and infrared data of several stages in the life cycle of the cloud vortices of a cyclonic storm near Kamchatka, Alaska.

Dash (−). The *dash* is used to indicate a change of thought or a change in sentence structure. It is also used for emphasis and to set off repetition or explanation. The dash may be used in place of parentheses when greater prominence to the subordinate expression is desired. Dashes should not be used with numbers because they might be mistaken for minus signs.

> The rectifier−that was the guilty component.
> I am suspicious of prognosticators−but I would not be surprised if tangible evidence of extraterrestrial, intelligent life will be found before the end of the present century.
> A traveling wave tube has been made with a helix fifteen-thousandths of an inch in inside diameter−about three times the diameter of a human hair.
> These molecules−formic acid, acetic acid, succinic acid, and glycine−are the very ones from which living things are constructed.

Data. *Data* is the plural of the Latin word *datum.* The singular form, *datum,* is rarely used in English. The word *data* means known, assumed, or conceded facts. Because the singular form is rarely used, the word *data* has become more and more acceptable, in all but the most formal English, as either singular or plural. Some writers treat *data* as a collective noun and use the singular verb with the word. For instance:

> The experiment's *data* was made available to the three laboratories.

However, there is a loud and armed camp of data pluralists who will split an infinitive with alacrity but will scream with outrage and scorn at any usage of a singular English verb with the Latin plural form. My fatherly advice is that it is always safe to use the plural unless you can afford a battle−and can afford to lose it, because you may well do so.

Division of Words. Break words only between syllables. When in doubt about syllabication, look the word up in the dictionary. Avoid breaking up a compound word that requires a hyphen in its spelling. Do not divide words of two syllables if the division comes after a single vowel: *among, along, atom, enough.*

Disinterested, Uninterested. *Disinterested* means fair, impartial, unbiased; *uninterested* means lacking interest or without curiosity. Although, in recent years, the two words have been used interchangeably in informal speech, your use of *disinterested* to mean lacking interest will raise many a reader's eyebrow.

Ditto Marks (″). Ditto marks are a convenience in tabulations and lists that repeat words from one line to the next, but ditto marks should not be used in correspondence.

Editorial "We." The substitution of *we* when *I,* obviously, is intended is considered by many writers as affected and pompous. However, in correspondence, this form is not only acceptable but is desirable when the writer is expressing his organization's policy or desires.

e.g. *e.g.* is an abbreviation of the Latin *exempli gratia,* meaning "for example." It is used to introduce parenthetical examples.

Ellipsis (. . .). A punctuation mark of three dots indicates that something has been omitted within quoted material. Four dots are used to indicate the end of a sentence. The *ellipsis* also is used to indicate that a statement has an unfinished quality.

Employee. This spelling is more common in the United States than *employe.*

Esquire, Esq. The form *Esquire* or *Esq.* is very formal. It is used after a man's name in the inside and outside address of a letter. Its usage is more prevalent in England than in the United States. The legal profession uses the form more frequently than other professions. If it is used, no other title—*Mr., Dr., Professor*—should precede the name it follows.

etc. *etc.* is an abbreviation for the Latin *et cetera,* which means "and so forth." It is sometimes used at the end of a list of items. Unless there is some reason for saving space, *etc.* should be avoided. An effective way to avoid this use is to introduce the list with *such as* or *for example.* The use of *and* with *etc.* is redundant.

Exclamation Mark (!). An *exclamation mark* (or *point)* is used after an emphatic interjection or forceful command. It may follow a complete sentence, phrase, or individual word.

Excuse, Pardon. Small trespasses are excused; greater faults and crimes are pardoned. Correspondence convention has made either word appropriate for perfunctory apologies.

Farther, Further. In informal speech, there is little distinction between the two words. In formal writing, *farther* applies to physical distances; *further* refers to degree or quantity.

> We drove sixty miles *farther* on.
> He questioned me *further.*
> The more he read about the matter the *further* confused he became.

Good, Well. *Good* is an adjective; *well* is both an adverb and adjective. One can say:

> I feel *good* (adjective).
> I feel *well* (adjective).

However, the meanings are different. *Good,* in the first sentence implies actual body sensations. *Well,* in the second sentence, refers to a condition of health—being "not ill." In nonstandard spoken English, *good* is sometimes substituted for *well.* For example:

> The centrifuge runs *good.*

Most educated persons would say:

The centrifuge runs *well.*

However. This adverb is a useful connection between sentences. *However* should be placed within a sentence rather than at the beginning.

A comparison of the two systems would be unfair; *however,* the differences are evident and interesting to observe.

Independent clauses connected by *however* are conventionally separated by a semicolon.

Hyphen (-). *Hyphens* are a controversial point in style. Modern tendency is to eliminate their use. Consult a good dictionary or the style manual of your organization. A *hyphen* is a symbol conveying the meaning that the end of a line has separated a word at an appropriate syllable, and that the syllables composed at the end of one line and the beginning of the next are one word. The *hyphen* is also used to convey the meaning that two or more words are made into one. The union may be ad hoc—for that single occasion—or permanent.

Light-yellow flame has a different meaning than light, yellow flame.

Hyphens are, therefore, used to form compound adjectival descriptive phrases preceding a noun: beta-ray spectrum, cell-like globule, 21-cm radiation, less-developed countries. Conventionally, *hyphens* are used to join parts of fractional and whole numbers written as words:

Thirty-three
One-fourth

Hyphens are used to set off prefixes and suffixes in differentiating between words spelled alike but having different meanings:

Recover his composure, and re-cover his losses.
Recount a story, re-count the proceeds.
Fruitless endeavor, fruit-less meal.

Hyphens are used between a prefix and a proper name:

pre-Sputnik, ex-professor
pro-relativistic quantum theory

Hyphens are used between a prefix ending in a vowel and a root word beginning with the same vowel:

re-elected
re-enter

Hyphenation has become more of a publisher's worry than a writer's. Most organizations set a style policy in hyphenation. Some follow the Style Manual of the University of Chicago or the Style Manual of the U. S. Government Printing Office. Some professional societies have style manuals for the writing done in their fields (e.g., the American Institute of Physics and the Conference of Biological Editors). Correspondence of professionals should follow the style preferred within their particular fields.

i.e. The use of *i.e.,* the abbreviation of the Latin *id est* (meaning *that is*), often saves time and space, but the English *that is* is preferable.

If, Whether. *If* is used to introduce a condition; *whether* is used in expressions of doubt.

> *If* the weather holds good, the space launch will be made.
> I wondered *whether* the space launch would be made.
> I asked *whether* the space launch would be made.

Imply, Infer. These two words are often confused. *Imply* means to suggest by word or manner. *Infer* means to draw a conclusion about the unknown on the basis of known facts.

In, Into, In To. *In* shows location; *into* shows direction.

> He remained *in* the laboratory.
> He came *into* the laboratory.

The construction *in to* is that of an adverb followed by a preposition.

> He went *in to* eat.

Its, It's. *Its* is a possessive form and means belonging to it. *It's* is the contraction for *it is.*

Lay, Lie. These verbs are often confused. *Lay* (principal parts: *lay, laid, have laid*) is a transitive verb meaning to put something down. *Lie* (principal parts: *lie, lay, have lain*) is an intransitive verb meaning to rest in a reclining position.

> I *laid* the tool on the bench.
> The old man *lay* resting on the sofa.

Leave, Let. *Leave* means to go away or to part with; *let* means to allow or permit.

Like, As. The preposition *like* is used correctly when it is followed by a noun or pronoun without a verb. *Like* should not be used as a conjunction; *as* should be used instead.

> He writes *like* an ignoramous.
> He looks *like* me.
> Winston tastes good *as* a cigarette should.
> Watch and do *as* I do.

May Be, Maybe. These two forms are often confused. *May be* is a compound verb; *maybe* is an adverb meaning perhaps.

Madam. *Madam* is a form of salutation used to address married and unmarried women in letters (the equivalent of *Sir* in addressing a man). The plural form is *Mesdames* or *Ladies* (the equivalent of *Gentlemen* or *Sirs*).

Dear Madam:
Mesdames: or
Ladies:

Dear Miss Jones or Dear Mrs. Smith is a construction more often preferred.

Messrs. *Messrs.* is the plural form for *Mr.*

Miss. *Miss* is the form for addressing an unmarried woman. The plural is *Misses.*

Mrs. The proper form for addressing a married woman is *Mrs.* In current practice, the signature line in a letter written by a woman includes the appropriate designation of Miss or Mrs. so that respondents may know how to address the writer in letters of reply. If the designation is not included, it is appropriate to use the address of *Miss.*

    (Mrs.) Katherine Biggs
    (Miss) Laura Jones

Misplaced Modifier. Modifiers should be placed near the words they modify.

    Wrong: Throw the horse over the fence some hay.
    Revised: Throw some hay over the fence to the horse.

Months. Write the names of months in full in the body of your letters. In the date line, you may abbreviate the names of months that have more than four letters if this will give the letter a neater appearance.

Money. Exact sums of money should be written in figures:

    25¢, $1.98, $10, $437.41, $83,655.

Round amounts are written in words:

    one hundred dollars, three thousand dollars, a million dollars.

None, No One. *None* is commonly used to refer to things (but not always); *no one* is used to refer to people. *None* may take either a singular or plural verb, depending on whether a singular or plural meaning is intended. *No one* always takes a singular verb.

Numerals. Practice varies in determining which numbers should be spelled out and which should be written as figures. Authorities frequently state rules that seem reasonable or preferable to them. A current trend is to use figures for all units of measure, such as yard, gram, meter, gallon, volt, acre, and bushel. Aggregate numbers are those resulting from the addition or enumeration of items. The tendency to differentiate between the two is rapidly disappearing. Most authorities recommend the spelling out of numbers under ten. Round numbers above a million are frequently written as a combination of figures and words:

    a 56 million dollar appropriation
    an employment force of 85 million

Most authorities recommend writing out a number at the beginning of the sentence. When two numbers are adjacent to each other, the first is spelled out:

six 120-watt lamps

The practice is to spell out small fractions when they are not part of a mathematical expression or when they are not combined with the unit of measure:

three-fourths of the area
one-third of the laboratory
one-half of the test population

However:

$1/2$ mile
$3/4$ inch pipe
$1/50$ horsepower
$1/4$ ton

Decimals are always written as figures. When a number begins with a decimal point, precede the decimal point with a zero as:

an axis of 0.52

Hours and minutes are written out except when A.M. or P.M. follow:

8:30 A.M. to 5:00 P.M.
six o'clock; nine-thirty; half past two
ten minutes for coffee breaks

Street numbers always appear in numerals:

1600 Pennsylvania Avenue, N. W.

When several figures appear in a sentence or paragraph they are written as numerals.

In Experiment 2, there was a total of 258 plants. These yielded 8,023 seeds—6,022 yellow, and 2,001 green.

Numbers indicating order—first, second, third, fourth, etc.—are called ordinal numbers. (Numbers used in counting—one, two, three, etc.—are cardinal numbers.) First, second, third, etc. can be both adjectives or adverbs. Therefore, the *ly* forms—firstly, secondly—are unnecessary and are now rarely used.

Oral, Verbal. These two words should not be confused. *Oral* refers to spoken communication; *verbal* has the general meaning of communication—written or spoken—which uses words.

Or. *Or* is a coordinating conjunction; it should connect words, phrases, or clauses of equal value.

Parallel Structure. Parallelism promotes balance, consistency, and understanding. Operationally, it means using similar grammatical structure in writing clauses, phrases, or words to express ideas or facts of equal value. A failure to maintain parallelism results in incomplete thoughts and illogical comparisons:

Faulty: Assembly lines poorly planned and which are not scheduled properly are inefficient.
Revised: Poorly planned and improperly scheduled assembly lines are inefficient.
Faulty: This is a group with technical training and acquainted with procedures.
Revised: This group has technical training and a knowledge of procedures.
Faulty: Before operating the boiler, the fireman should both check the water level and he should be sure about the draft.
Revised: Before operating the boiler, the fireman should check both the water level and the draft.

Parentheses ( ). Parentheses are used to enclose additional, explanatory, or supplementary matter to help the reader in understanding the thought being conveyed. These additions are likely to be definitions, illustrations, or further information added for good measure. *Parentheses* are used also to enclose numbers of letters to mark items in a listing or enumeration.

Per. *Per* is a preposition borrowed from the Latin, meaning *by, by means of, through,* and is used with Latin phrases that have found their way and use in English:

per diem, per capita, per cent, per annum

Per has established itself by long range usage in business English.

Per Cent. *Per cent* is two words, but more and more the phrase is being written as one word. The sign, %, is convenient and appropriate in tables, but its use should be avoided in text.

Period. *Periods* are used to mark the end of a complete declarative sentence and abbreviations. *Periods* are also used in a request, phrased as a question out of courtesy:

Won't you let me know if I can be of further service.

Phenomenon, Phenomena. The plural form *phenomena* is frequently misused for the singular *phenomenon.*

Prepositions. A *preposition* is a word of relation. It connects a noun, pronoun, or noun phrase to another element of the sentence. Certain prepositions are used idiomatically with certain words. For example:

Knowledge of
Interest in
Hindrance to
Agrees with (a person agrees to) a suggestion, agrees with (another)
Agrees in (principle to) a suggestion
Agrees to (a plan)

Obedience to
Responsibility for

- Years ago, it was fashionable for grammarians to stigmatize prepositions standing at the end of sentences. Actually, it is a characteristic of English idiom to postpone the preposition. Most attempts to avoid prepositions at the end of a sentence lead to unfortunate blunders and heavy handed, awkward sentences.

Principal, Principle. *Principal* is used both as a noun and an adjective. As a noun it has two meanings: (1) the chief person or leader; and (2) a sum of money drawing interest. As an adjective, *principal* means "of main importance" or "of highest rank or authority." *Principle* is always a noun and means fundamental truth or doctrine, or the basic ideas, motives or morals inherent in a person, group, or philosophy.

Punctuation. *Punctuation* is one means by which a writer can achieve clarity and exactness of meaning. Sloppy punctuation can distort meaning and confuse the reader. There are two principles governing the use of punctuation marks:

1. The choice and placement of punctuation marks are governed by the writer's intention of meaning.
2. A punctuation mark should be omitted if it does not clarify the thought.

There is a modern tendency to use open punctuation marks—that is, to omit all marks except those absolutely indispensible. A common mistake of the inexperienced writer is to overuse punctuation marks, especially the comma. A good working rule is to use only those marks for which there is a definite reason, either in making the meaning clear or in meeting some conventional demand of correspondence. To illustrate, notice the difference in meaning between the following two sentences:

The professor said the student is a fool.
"The professor," said the student, "is a fool!"

Question Mark (?). This punctuation mark is used to denote the end of a question. The *question mark* is placed inside a quotation mark only when the quoted matter itself is a question.

Quotation Marks ("). *Quotation marks* are used to enclose direct quotations, titles of articles and reports, and coined or special words or phrases.

Said. In legal documents, the use of *said* as a demonstrative pronoun (this, that, these, those) has a long tradition of use (e.g., *said* Mr. Peterson, *said* dwelling). However, in business correspondence, it should be avoided as a cliché.

Semicolon (;). The principal use of the *semicolon* is to separate independent clauses that are not joined by a conjunction or are joined by a conjunctive adverb or some other transitional term (e.g., *therefore, however, for example,* *in other words*). A semicolon is also used to separate clauses and phrases in a series when they already contain commas.

> Among the articles offered for sale were a harpsichord, which was at least 200 years old; a desk, which had the earmarks of beautiful craftsmanship; and a table and four chairs, which were also antique.

Slow, Slowly. *Slow* is both an adjective and an adverb; *slowly* is an adverb. As adverbs, the two forms are interchangeable, but *slow* is more forceful than *slowly*.

Split Infinitives. Usage has won out. To split an infinitive is no longer viewed as a grammatical misdemeanor. Frequently the interpolation of a word between the parts of an infinitive adds clarity and emphasis. For example:

> To carefully examine the evidence
> To forcefully impeach
> To seriously doubt
> To better equip

Should, Would. *Should* and *would* are used in statements that suggest some doubt or uncertainty about the statement that is being made. Many years ago, *should* was restricted to the first person, but usage now is so divided in the choice of these words that personal preference rather than rule is the guide today. However, consistency in usage in the same letter should be followed.

> I would greatly appreciate your granting me an interview.

or:

> I should greatly appreciate your granting me an interview.

Than, Then. *Then* is an adverb relating to time; *than* is a conjunction in clauses of comparison.

> *Then* came the dawn.
> Fortran is a more universal programming language *than* are any machine languages.

There Is, There Are. These expressions delay the occurrence of the subject in a sentence. The verb in the expression must agree in number with the real subject.

> *There is* a difficult problem associated with the system.
> *There are* two solutions to the problem.

Often these expressions can be omitted with no loss to the sentence.

> A difficult problem is associated with the system.
> Two solutions are available for the problem.

or:

> The problem has two solutions.

Toward, Towards. The two forms are interchangeable.

Underlining. In correspondence, *underlining* is used (in place of italics):
1. To indicate titles of books and periodicals.
2. For emphasis:
   Words that would be heavily stressed when spoken are *underlined*.
3. To indicate foreign words.

Unique. The word means single in kind or excellence, unequalled; therefore, it cannot be compared. It is illogical to speak of (or to write): a more *unique* or most *unique* thing.

Very. *Very* is an intensive word. It has become so overused that its intensive force is slight. If you are tempted to use *very* in a sentence or in the complimentary close, ask yourself what the word adds to the meaning you wish to convey. If the answer is negative, do not use it.

Which. *Which* is a pronominal word, used in both singular and plural instructions. It refers to things and to groups of people regarded impersonally (The crowd, which was a large ...).

Who, Whom. *Who* is the form used in the nominative case (as the subject) and *whom* is the form used in the objective case (when the form is the object of a verb or of a proposition).

He is the one *who* is responsible.
For *whom* the bell tolls.
Whom did he ask to serve as secretary?

Your, You're. *Your* is a possessive pronoun.

Your company.

You're is a contraction for *you are.*

You're to leave before noon.

A SELECTED BIBLIOGRAPHY

Acronyms Dictionary: A Guide to Alphabetical Designations, Contractions, and Initialism, Detroit, Mich.: Gale Research Co., 1960.
Chambers Technical Dictionary, 3rd ed., New York: The Macmillan Company, 1958.
Condensed Chemical Dictionary, 6th ed., New York: Reinhold Company, 1961.
Evans, Bergen, and Cornelia Evans, *A Dictionary of Contemporary English Usage,* New York: Random House, 1957.
Gilman, William, *The Language of Science,* New York: Harcourt, Brace, and World, 1961.
Kapp, Reginald O., *Presentation of Technical Information,* New York: The Macmillan Company, 1957.
Mandel, Siegfried, and David L. Caldwell, *Proposal and Inquiry Writing,* New York: The Macmillan Company, 1962.
A Manual of Style, Chicago: University of Chicago Press, 1949.
Menning, J. H., and C. W. Wilkinson, *Writing Business Letters,* revised ed., Homewood, Ill.: Richard D. Irwin, Inc., 1959.
Nicholson, Margaret, *A Dictionary of American-English Usage Based on Fowler's Modern English Usage,* New York: A Signet Book Published by the New American Library, 1958.
Perrin, Porter G., *Writer's Guide and Index to English,* 4th ed., Chicago: Scott, Foresman and Company, 1965.
Rathbone, Robert R., *Communicating Technical Information,* Reading, Mass.: Addison-Wesley Publishing Company, 1966.
Readings in Communication from *Fortune,* edited by Francis William Weeks, New York: Holt, Rinehart, and Winston, 1961.
Redding, Charles W., and George A. Sanborn, *Business and Industrial Communication:* A Source Book, New York: Harper & Row, 1964.
Sheppard, Mona, *Plain Letters,* New York: Simon and Shuster, Inc., 1960.
Singer, T. E. R., Editor, *Information and Communication Practice in Industry,* New York: Reinhold Publishing Corporation, 1958.

Appendix

Smart, Walter Kay, Louis William McKelvey, and Richard Conrad Gerfen, *Business Letters,* 4th ed., New York: Harper & Brothers, 1957.

Steen, Edwin B., *Dictionary of Abbreviations in Medicine and the Related Sciences,* 2nd ed., London: Cassel, 1963.

Strunk, W., *The Elements of Style,* New York: The Macmillan Company, 1959.

Tichy, Henrietta J., *Effective Writing for Engineers Managers, Scientists,* New York: Wiley, 1966.

United States Government Printing Office Style Manual, Washington, D. C.; U. S. Government Printing Office, Revised Edition, January 1967.

Websters Seventh New Collegiate Dictionary, Springfield, Mass.: G. & C. Merriam Co., 1963.

Weisman, Herman M., *Basic Technical Writing,* Columbus, Ohio: Charles E. Merrill Books, Inc., 1962.

INDEX

Abbreviations, 45
Acceptance letters, 154, 157
Accepting position letter, 149
Acknowledgment of appreciation letter, 160
Acknowledgments of orders, 80, 86–87
  organization of acknowledgment letters, 86
Acknowledgments to application letters, 141 ff
Adjustment letters, 110–113
Agenda, the, 181–183
American Chemical Society, 187
Applied Physics Letters, 167
Appreciation letters, 158
Attention line, 44, 49–50
Authorization letter, 118–120
Automated methods, in correspondence, 5

Bacon, Francis, vii, 4
Becker, Joseph D., 175
Beginnings of job application letters, 126 ff
Block style in letters, 34, 36, 40
Body of the letter, 45
Business correspondence, 23

Carbon copy notations, 47
Cherry, Colin, 10
Claims letters, 106–108
  humor in, 108–109
  organization of, 107
Clarity, 18
Clichés, 19–20
Closing paragraph, 30, 31
  effective methods of, 32
Coherence in letters, 21
Communication, definition of, 7
  feedback, 9
  how it works, 8–10

Communication, ingredients of, 8
  meaning in, 10–11
  process of, 7
Communications to the editor, 166
Complaint letters, 106
Complimentary close, 45
Conciseness, 19
Condolence letters, 161
Congratulations, letters of, 160–161
Copernicus, 4
Correspondence, applying principles, 55 ff
  history, 4
  mechanics, 33 ff
  as a medium, 11
  planning, 24 ff
  principles and fundamentals, 1 ff
  psychology, 12 ff
  role of, 5
  style, 16–22
Correspondence between professionals, 162 ff
Courtesy in letters, 20
Cusack, N. E., 167

Date line, 41
Declination letters, 154, 157
Directness, 19

Editor's mail column, 166, 169 ff
Electronic Letters, 167
Employment letters, 122 ff
  acceptance, 149
  acknowledgments, 141
  features for effectiveness, 123
  follow-up, 141, 143–145
  job application, 125 ff
  recommendation, 146
  reference, 145
  refusing a job, 150

216 Index

Employment letters, resignation, 149
Enclosures, 47
Envelope, 48–50

Faults in clarity and precision, 18–19
Feedback, 8, 9, 11, 12
File number, 43
Follow-up letters to applications, 141, 143–145
"Form 57" in government employment, 131
Form replies to requests, 65–67
Full block, 34, 36, 40

Galileo, 4
Grammar, 193 ff

Hanging indentation style, 39, 40
Heading of letter, 40–41
House of Fugger, 4

Identification line, 46
Indented style, 40
Industrial Research, 23
Information explosion, 162
Inquiry letters, 57 ff
  organization of unsolicited inquiry letters, 59–60
  solicited, 57–59
  unsolicited, 57, 59–64
Inside address, 41–42
Introduction, letters of, 153–154
Invitation, letter of, 154
  to get together, 156
  to participate in a committee, 155
  to speak, 155

Job analysis in writing application letter, 124–125
Job application letter, 125 ff
  adapting your qualifications, 129–130
  components, 126 ff
  resume, 131 ff
  securing action, 130–131
Johnson, Samuel, 7
Journal of the American Chemical Society, 166
Journal of Applied Physics, 166
Journal of Biological Chemistry, 167
Journal of Chemical Physics, 166
Journal of Colloid and Interface Science, 166

Journal of Geophysical Research, 167

Kepler, 4

Lawrence, Ernest, 162, 163–166
Layout of letters, 34 ff
Letter, the, 12 ff
  defined, 13
  planning, 24 ff
  qualities of, 13–15
  steps in writing, 27 ff
    determining purpose, 27
    gathering information, 28
    outlining and organizing, 28–29
    reader consideration, 28
    writing, 29
Letterhead stationery, 34
Letter proposal, the, 103–105
Letters to the editor, 166 ff
Lewis, Ralph W., 171

Marwaha, A. S., 167
Meaning, 9, 10–11
Mechanics of correspondence, 33 ff
  carbon copies, 47
  envelope, the, 48
  framing the letter on the page, 34
  mechanical details, 40 ff
    components of the letter, 40 ff
  memorandum, 50 ff
  post script, 48
  second and succeeding pages, 48
  standard styles, 34–40
  stationery, 33
Meetings, 181
Memoranda, professional, 162 ff
Memorandum, the, 50–53, 176 ff
Memorandum for file, 179–180
Miller, Samuel E., 172
Minutes of meetings, 183 ff
Modified block, 34, 35, 40

NOMA style in letters, 34, 38, 40
Notices of meetings, 181–182
Novel beginnings in application letters, 126–127

Objectivity in scientific writing, 22–23
On Human Communication, 10
Opening paragraph, 30, 31

Index

Order letters, 80–85
  organization of, 81
Outline of letter, 28–29

Paragraph, 22
Personal correspondence, 152 ff
  acceptance, 154, 157
  acknowledgments of appreciation, 160
  appreciation, 158
  condolence, 161
  congratulation, 160–161
  declination, 154, 157
  introductions, 153–154
  invitations, 154–157
  reservations, 153
  style in, 152
Personal data sheets, 131 ff
Personal notations on envelopes, 50
Physical Review Letters, 167
Physics Letters, 167
Physics Today, 23, 163
Postscript, 48
Precision in writing, 18
Proceedings of the American Mathematical Society, 166
Proceedings of the IEEE, 166
Professional letters, 162 ff
Proposal letters, 88, 98 ff
  elements of, 101
    introduction, 101
    technical description, 101
    technical presentation, 101
    transmittal letters for, 102–103
Punctuation, 43, 193 ff

Question beginnings in application letters, 128–129
Quotation letters, 80, 87

Reader, the, 15–16
Recommendation letters, 146–148
Reference beginnings in application letters, 127–128
Reference letters, 145
Reference lines, 43, 44
Refusing a job letter, 150–151
Replies to inquiries and requests, 65–80
Reports, 12–13
  differences between reports and letters, 13

Request for bid letters, 99–100
Reservation letters, 153
Resignation letter, 149–150
Resume, 131 ff
Rokeach, Milton, 173
Rutherford, Ernest, 162, 163–166

Sachs, Mendel, 175
Sales letters, 88 ff
  beginnings in, 91–93
  convincing the reader, 94
  creating interest, 93–94
  definition of, 88
  examples of, 95 ff
  motivating the reader, 94
  role of, 89–90
  steps in construction of, 90
Sample outline of letter, 29
Science, 23, 167, 175
Science Technology, 23
Second and succeeding pages, format in letters, 48
Self-appraisal in writing application letter, 124–125
Semantics, 10–11
Semiblock, 34, 35, 40
Sentence defined, 21
Signature, 46
Simplified style in letters, 34, 37, 40
Sincerity in letters, 20
Solicited inquiry letters, 57–59
Soule, George H., 170
Special purpose letters, 106 ff.
  adjustment, 110 ff
  authorization, 118–120
  claims, 106–108
  complaint, 106
  instructions, 113–118
  transmittal, 120–121
Stanerson, B. R., 190
Stationery considerations, 33–34
Style, 16 ff
  clarity, 18
  conciseness, 19
  conciseness and directness, 19
  courtesy, 20
  defined, 17
  directness, 19
  precision, 18, 19
  sincerity, 20

Subject lines, 43, 44
Summary beginnings in application letters, 128
Symbols, 10

Technical products, 89
Thesis statement, 24
Transmittal letter, 120–121

Unity in letters, 21

Unsolicited inquiry letters, 57, 59–64
Usage, 193 ff

Wiener, Norbert, 8
Words, 10

"You" psychology, 13–15, 126

Zip code, 49–50